Spirals

Spirals

From Theodorus to Chaos

Philip J. Davis
Division of Applied Mathematics
Brown University

with contributions by

Walter Gautschi
Department of Computer Sciences
Purdue University

and

Arieh Iserles
Division of Applied Mathematics and
Theoretical Physics
University of Cambridge

A K Peters
Wellesley, Massachusetts

Editorial, Sales, and Customer Service Office

A K Peters, Ltd.
289 Linden Street
Wellesley, MA 02181

Library of Congress Cataloging-in-Publication Data

Davis, Philip J., 1923–
    Spirals : from Theodorus to chaos / Philip J. Davis ; with
contributions by Walter Gautschi and Arieh Iserles.
      p. cm.
    Includes bibliographical references and index.
    ISBN 1-56881-010-5
    1. Spirals. I. Gautschi, Walter. II. Iserles, A. III. Title.
QA567.D38 1992
516′.15–dc20   93-16735
                                               CIP

Printed in the United States of America
97 96 95 94 93   10 9 8 7 6 5 4 3 2 1

To the Memory of

Victor, Lord Rothschild

L'esprit humain fait toujours de progrès, mais ce progrès est spirale.

—Madame de Staël

Contents

Acknowledgements

I wish to acknowledge conversations and correspondence with the following individuals, who, in different ways, were of great help to me in the preparation of these lectures.

Jeff Achter, Douglas E. Baker, Christa Binder, Fred Bisshopp, David Busath, Wai-fong Chuan, Constantine Dafermos, Frank B. Davis, Marguerite Dorian, Walter Gautschi, David Gottlieb, Branko Grünbaum, Phyllis Hersh, Reuben Hersh, Edmund Hlawka, Arieh Iserles, Paris Kanellakis, Michael Katz, Candace Kent, Douglas S. Klein, Hüseyin Koçak, Jeffery Leader, Kevin Manbeck, Kenneth Miller, George Phillips, Peter Renz, Peter Schmidt, Frank Stenger, Trevor Stuart, Gerald Toomer, Andries Van Dam, Roselyn Winterbottom.

I also wish to acknowledge the support of the Alfred P. Sloan Foundation.

Part I
The Hedrick Lectures

Foreword to
The Hedrick Lectures

*Concerning this shell in whose shape I think I can discern
... the work of some hand not acting "at random," I ask my-
self: Who made it?*

    *But soon my question undergoes a transformation. It takes
a short step forward along the path of my naiveté and I begin
to inquire by what sign we recognize that a given object is or
is not made by a man?*

—Paul Valéry, *Man and the Sea Shell*

The Hedrick Lectures, presented here together with amplifica-
tions, were delivered on the occasion of the seventy-fifth anniversary of the Mathematical Association of America (MAA).
The lecture series itself, named after Earl Raymond Hedrick,
the first president of the MAA, has been a central feature of
the association's annual summer meetings for a long time.

Over the past three quarters of a century, the MAA has played
a tremendous role in supporting mathematical education, publication, research and in generating mathematical enthusiasm
in the United States. Our inheritance from the past activities

of the MAA is substantial. We are all heirs to the traditions it
has fostered, and we wish the MAA well as it speeds forward
toward its century mark.

Anniversaries of whatever kind put one in mind of the passing years, and a mathematical anniversary serves to remind us
that mathematics has been pursued – often with passion[1] – for
four thousand years at the very least. Its history and its poten-
tialities admit of no ethnic, nationalistic, or gender boundaries.
Mathematics is coexistent and coextensive with civilization itself. It has in the past and can in the future act as a powerful
binder between varieties of ethnic sensibilities.

The study of mathematical history displays an arena wherein
group assent must be sought and obtained and yet where individual genius is of critical importance. As William James wrote:
"...without individual genius, the community stagnates; without the community, individual genius has no arena."

The study of mathematical history leads us also to the recognition that mathematics has not remained constant as regards
its inner or outer goals or as regards its philosophical interpretations or orientations. Very often, in the words of Justice
Oliver Wendell Holmes (in another context), mathematics "lays
its course by a star it has never seen. ... " In so doing, and over
such an extended period of time, its material can also become
Time's Exile in that what were once thought to be major ac-
complishments, insights and connections are found cast into the
dust of irrelevance.

In keeping, then, with the spirit of this anniversary, I have
planned this book so as to exhibit a variety of things: some history, some philosophy, some anecdotes, a fair amount of mathematics, naturally; some things old, some things absolutely new,
some things proved, and many things that invite exploration
at a variety of levels of sophistication; many things that invite
discussion, conjecture and proof. The lectures and the contemporary supplements have been organized loosely around one
central theme: the study of a certain difference equation that
I have called by the name of an ancient Greek mathematician:
Theodorus of Cyrene. I shall look at this difference equation

in the light of mathematical concerns that have grown and changed over the past twenty-five hundred years.

I have occasionally taken the liberty of free associating on some of the ideas. These rambles – often tangential – have been placed in the notes at the end, and the reader's attention is directed to them. To me, mathematics has always been more than its form, or its content, its logic, its strategies, or its applications. Mathematics is one of the greatest of human intellectual experiences, and as such merits and requires a rather liberal approach. But I hope I have been able to moderate my natural effusiveness so that it won't be said of me, as it was of George Eliot's Casaubon, that "he dreams footnotes."

Lecture I

What is a Spiral? Spirals Old and New

The inspiration for these lectures comes from a paper by Edmund Hlawka, of the Technical University in Vienna, on a certain discrete spiral that in German has been called the *Quadratwurzelschnecke* [Hlawka 1980]. Translated literally: the square root snail.[2] (See Historical Supplement H.)

I, personally, have always thought of a *schnecke* as one of those little twisted pastries one finds in European-style bake shops. In a mathematical age that has discussed Strudelpunkte, pretzel universes, bakers' transformations, blue bagel chaostrophes and bifurcations, and considering that in the book on my shelf next to where these things are discussed, there is a mathematical "delicatessen" where the rigorous concept of Wiener sausages is cooked up and digested, it is by no means absurd to take inspiration from square root continental pastries.

The statement in Plato's *Theaetetus* that Theodorus of Cyrene discussed the irrationality of $\sqrt{2}, \sqrt{3}, \ldots$, and stopped at $\sqrt{17}$, has, over the millennia, evoked much speculation.[3] Why did Theodorus stop at $\sqrt{17}$? This speculation continues to this day. The interested reader will find numerous answers in the literature.[4]

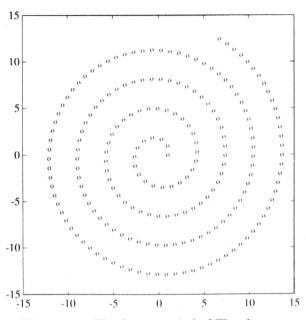

√0̄=0
origin?

Figure 1:   The discrete spiral of Theodorus.

An answer of considerable fantasy was provided seventy years
ago by a certain J. H. Anderhub, an inspired mathematical ama-
teur [Anderhub 1941]. Anderhub imagined that Theodorus constructed $\sqrt{2}, \sqrt{3}, \ldots$ by a sequence of contiguous right-angled
triangles. In each triangle, each outer leg is of length 1. Anderhub observed that the resulting snaillike figure is such that
$\sqrt{17}$ arises from the last triangle for which the total figure is
non-self-overlapping.

If one goes beyond to $\sqrt{18}, \sqrt{19}, \ldots$, the figure overlaps itself
and would be quite messy to draw in the sandboxes in which
mathematical myth claims all theorems of ancient Greek geometry were drawn. Ergo: he stopped at $\sqrt{17}$. (See figs. 1, 2.)

Though Anderhub's "solution" is interesting, it has little historic cogency or plausibility. Anderhub's idea was embedded by
the present author in a light literary *jeu d'esprit*.[5] Nevertheless,
the figure (when extended indefinitely) has recently attracted
the attention of a number of mathematicians who have raised

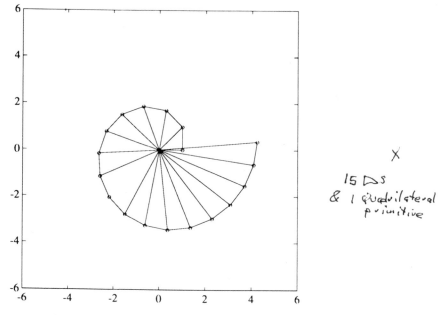

Figure 2:   The spiral of Theodorus exhibiting the "solution" of Anderhub.

contemporary questions concerning it. Thus, E. Hlawka proved (W. Neiss [1966], originally) that if (r_n, θ_n) are the polar coordinates of the vertices of the spiral, then the angles θ_n are equidistributed mod (2π) in the sense of Hermann Weyl. He did this essentially in the following steps:

A careful analysis using the Euler–Maclaurin formula[6] yields

$$\theta_n = 2\sqrt{n} + K + (1/6)(1/\sqrt{n}) + (1/4)5^4(n-1)^{-(3/2)} + \cdots,$$

for a certain constant K.

Since

$$r_n = \sqrt{n},$$

this shows, firstly, that the spiral that emerges from this geometric process is asymptotic to a spiral of Archimedes. (See fig. 4.) Secondly, by making use of basic results in the theory of

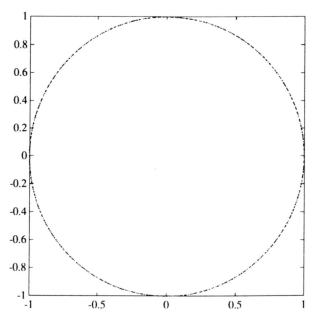

Figure 3:   The first thousand angles of the spiral of Theodorus exhibiting the equidistribution property.

equidistribution, in particular, a theorem of Fejér,[7] we conclude first that $(\sqrt{n})$, that is, the fractional part of $\sqrt{n}$, and then θ_n are equidistributed. (See fig. 3.)

I think that, using this approach, we can also show that the vertex angles are equidistributed integerwise, in the sense of Ivan Niven.[8]

Hlawka also demonstrated (E. Teuffel [1958] originally, and also H. Rindler) that the angles θ_n are algebraically independent in the sense that no two hypothenuses, when extended, will coincide, and that when the sides of length 1 are extended indefinitely, they will not pass through any of the other vertices of the total figure. Analytically, the first reduces to showing the impossibility of satisfying

$$\theta_{n+L} - \theta_n = g\pi$$

in positive integers g, n, and L.

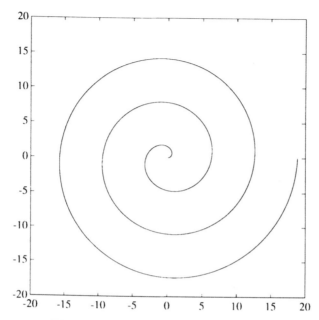

Figure 4:   The spiral of Archimedes.

The proof of Hlawka draws on the circle of ideas of the so-
called postulate of Bertrand–Tschebyscheff that for $n > 1$, there
is always a prime between n and $2n$, and from a theorem of
Besicovitch that implies that if p_k are the consecutive primes,
then $\sqrt{p_1}, \ldots, \sqrt{p_s}$ are independent over the rationals.[9]

primes

I should like to generalize Figure 1, raising some questions as
I go along (and answering a few). Some of these questions were
inspired by the hard facts of computation.

But first, I would like to talk about a simple idea: that of a
spiral. In our college courses, we do not teach too much about
spirals. Come to think of it, we might even say: What *is* there
to teach ? Isn't a spiral just an exercise in the first-year calculus
book? And if one looks in the indexes of mathematical textbooks
or monographs, one does not come across the notion too often.

In preparation for these talks, I interrogated the silver disc
(marketed by the Silver Platter Co.) that contains the Mathematical Reviews of the past fifty years. I turned up about two

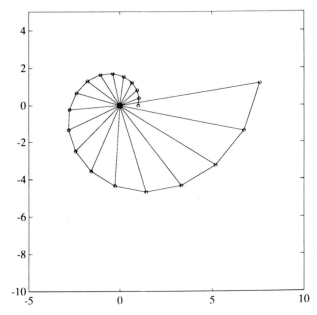

Figure 5:   Spiral of Bernoulli with chambered nautilus
parameters.

hundred fifty entries under spirals and an equal number under
a variety of related designations. Most of the citations relate to
physics. Many relate to conformal mapping.

I found Pythagorean spirals. I found lambda-spirals; I found
spiral galaxies; spiral waves. There is the marriage of spirals
and convexity theory considered by Bourgin and Renz [1989].
There is the marriage of spirals and quasiconformal mappings
considered by Gehring [1978]. There are the spirals generated by
Diophantine Gauss sums that have been considered by Coutsias
and Kazarinoff [1987].

There are spirallike analytic functions, i.e., functions that
map the unit disc into something that contains a logarithmic
spiral. There are spirals in Hilbert space considered by Wiener
and later by Kolmogoroff in one of the most influential papers in
modern probability theory, and they play an interesting role in

Figure 6:   Inscription on the Bernoulli monument.

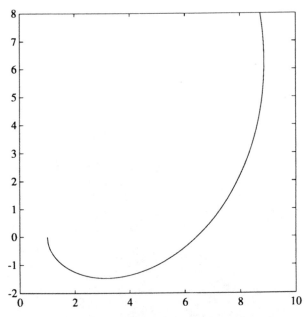

Figure 7: The spiral of Norwich.

second-order stochastic processes. There are spirals in Hilbert space considered by von Neumann and Schoenberg [1941].[10]

There are the spirals or the helices of genetics that make us what we are and to some extent what we will be. There are the spirals or helices in the internal structure of a tornado, spirals that can whish us off, like Dorothy, to the Land of Oz.[11] There are the helix-to-random coil transitions of polymer physics in which configurations pass from minimum energy to maximum entropy and which provide us with plastics from bank cards to heart-valve replacements. The arondissements of Paris are numbered spirally, beginning at the Louvre. And I knew independently, a fact not yet on the silver disc, that there had been a recent conference on spirals.[12]

I was in culture shock. So much interesting material on such a simple idea. And I was in shock not only from information overdose, but since the silver disc created information on the terminal screen in twenty colors, alternately flashing and

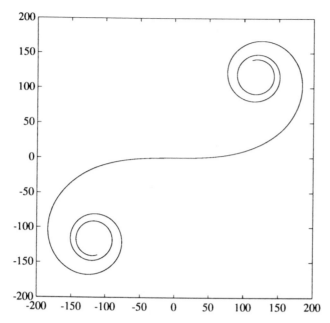

Figure 8:   The clothoid or spiral of Cornu.

alternately in polka dots, I had the impression I was standing on the Strip at Las Vegas and not in the Science Library of Brown University. It would seem that the new generation cannot think properly except in color-coded overlays. Marvellous! One can imagine educational apologetics fifty years from now: Why should one study the art of polychromatic graphical layout? Answer: It teaches one to think properly. Doesn't that kind of argument sound familiar?

The famous ex-Soviet author and poet, Joseph Brodsky, recently gave a talk at one of the publishing fairs in Germany: How on earth can one deal with the fifty thousand books that are published annually? Brodsky's answer was: Convert the ideas to poetry. Poetry is condensed thought and emotion.

The standard answer in mathematics when one is snowed by a blizzard of material is to align oneself with a Center (in the global sense) and then ignore all non-Center work. My answer to how I could deal with the hundreds of papers on spirals that

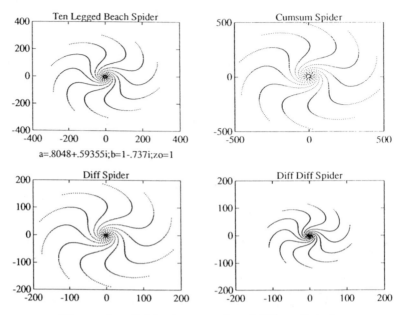

Figure 9: Spider integrated and differentiated.

the silver disc highlighted may not be such a good one, but it worked for me: Get in touch with a few good friends and tap them.

Though Plato remarks in the *Timaeus* that spirals occur among the planetary motions (see Historical Supplement A), he does not give any mathematical details; let me therefore start with what is probably the first spiral to be treated mathematically: the spiral of Archimedes. Archimedes[13] wrote a book about this spiral: the *Peri Elikon*. In classical Greek, an 'elix is a winding, a coil, a curl, a bracelet, or an earring. And, of course, a twisted Danish pastry. (See fig. 4.)

Archimedes defined his spiral by a ray rotating uniformly coupled with a point that moves uniformly on that ray. This may have been the first instance in mathematics of defining a curve as the result of two independent uniform motions. Archimedes put his spiral to the use of trisecting the angle and squaring the circle. The deepest result of Archimedes, apparently,

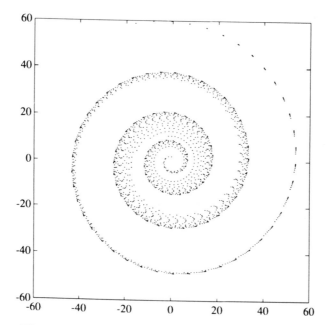

Figure 10:   Kronecker (Theodorus, Theodorus).

is his determination of the area swept out by a spiral after one rotation.[14]

Now the spiral of Archimedes may not be due to Archimedes but to a certain Konon of Samos![15] Who was this Konon of Samos? He was Court Astronomer (and very likely, Court Astrologer) to King Ptolemy Euergetes of Egypt. He was a friend of Archimedes who thought highly of him. After travelling in the western portions of the Greek world, in search of astronomical and meteoric observations, he settled in Alexandria. He researched solar eclipses and participated in the development of the Greek astronomical calendar. He wrote a book on the mutual contact of conic sections. Konon died young. Alas. Like Galois; like Urysohn or Paley; like Lippman Lipkin in St. Petersburg or Arthur Buchheim at the Manchester Grammar School, and of whom very few have now heard; like many others.[16]   ](?)

Early in his marriage, Ptolemy had to lead his army in a war against the Seleucids in Syria. (This was in 246 B.C. at the

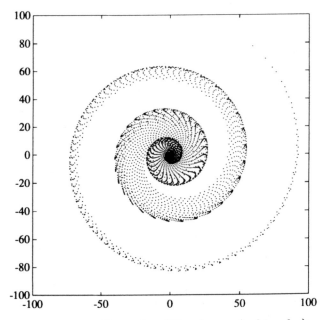

Figure 11a: Kronecker (Theodorus, Archimedes).

very height of the Ptolemaic dynasty.) His wife Berenice was advised to sacrifice a lock of her hair to the Temple of Arsinoë Zephyritis as an insurance policy for his safe return. She did so, and King Ptolemy came back to his wife hale and hearty. But woe! Shortly thereafter, the lock disappeared from the temple. And the soothsayers naturally interpreted this as a bad omen. Thereupon, Konon calmed her down by saying that he had discovered her lock (I conjecture it was a spiral coil) in a new constellation between Leo, Virgo, and Boötes! (On today's star maps: Coma Berenices. In Greek: Berenikes Plokamos).[17]

This pretty story was told originally by Callimachus and reworked much later by Catullus. Legend? Possibly; but there is nothing in it that is unbelievable.

But what is a spiral?

Of course, there are the specific instances of spirals. There is the spiral of Bernoulli (see fig. 5), or of Descartes, or of Gregory, or of Torricelli, depending upon whose drummer you march by,

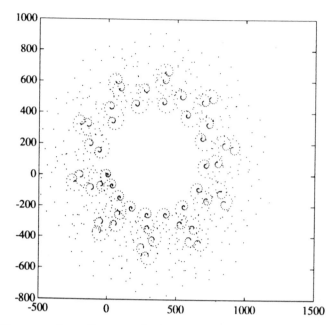

Figure 11b: Cumsum (Kronecker (Marigold, Bernoulli)).

and it is also referred to as the logarithmic, the exponential, or the equiangular spiral.[18] In a certain sense, probably because of the role it plays in the theory of linear differential equations and because of its seeming omnipresence in nature, this is the spiral *par excellence.*[19]

This, also, is the curve designated by Jacob Bernoulli in 1694 as the *spira mirabilis*, the wonderful spiral, wonderful in virtue of its numerous self-reproducing properties, which he took as a symbol for a variety of self-reproducing aspects of the natural and theological worlds.[20] This was the spiral honored by him with the motto *Eadem mutata resurgo* ("Though changed, I rise again the same"). This is the spiral that is carved on his gravestone, together with the motto quoted either, as some have conjectured, to assert his belief in the resurrection of the body or to assert his fervent hope in the same.[21]

There are the spirals $r = a\theta^n$, which, for $n = -1$ is the hyperbolic spiral (Varignon, 1704), for $n = -\frac{1}{2}$ is the lituus

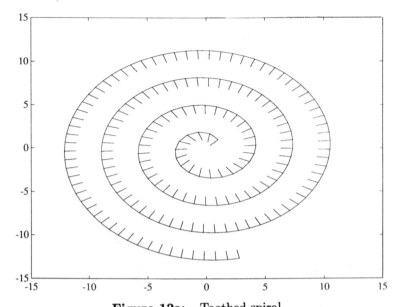

Figure 12a: Toothed spiral.
j = iteration number, $d = \text{rem}\,(j, 3)$; $z = z + i^d z/|z|$.

(Cotes, 1722), for $n = \frac{1}{2}$ is the spiral of Fermat (1636), and for $n = 1$ is the spiral of Archimedes.

There are the sinusoidal spirals, $r^n = a^n$ (cos or sin) $n\theta$, which go back at least as far as Maclaurin in 1718.

There is the Cotes spiral, which is the path of a particle moving under the inverse cube law of attraction.[22]

There is the spiral of Norwich, so named by J. J. Sylvester because of a meeting of the British Association that took place in Norwich in 1868. (See fig. 7.) It is defined by: the radius of curvature equals the length of the radius vector. This spiral coincides with the spiral of Sturm, and it turns out that Jacopo Riccati had done the general theory of this kind of thing years before Sylvester.[23] (See fig. 8.)

There is the marvellous spiral of Cornu, which I once dubbed "the most beautiful of the mathematical curves."[24]

There are spirals on a cone and spirals on the surface of a sphere, ...; and so it goes, on and on. This lecturer cannot

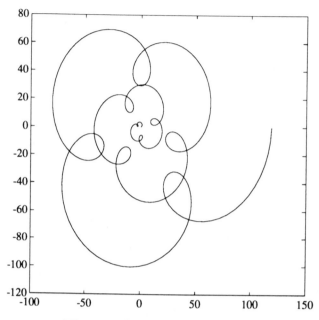

Figure 12b: Is this a spiral?

undertake to name all the special spirals that have been studied
over the centuries; and in the second lecture, he is about to add
one more to the list.[25]

But what is a spiral, generally speaking? (See figs. 12–22.)
The dictionaries, mathematical or otherwise, aren't much help.
They give definitions for which counterexamples are easily pro-
vided.

Encyclopaedia Britannica, 11th ed.: A spiral is a curve that
winds around a fixed point.[26]

American Heritage Dictionary: Locus of a point moving around
a fixed center at a monotonically increasing or decreasing distance.

An old scientific dictionary in my library, whose cover was
ripped off: Spiral – a term used generally to describe any geometric entity that winds about a central point while also receding from it.

James and James' Mathematical Dictionary: Spiral – No entry.

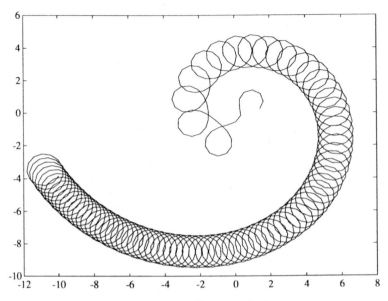

Figure 13: Spring spiral.

$$d = .5\exp(.5ij); z = z + dz/|z|.$$

What is a discrete spiral? When can an infinite sequence of points in the plane or in higher dimensions be organized into a spiral? (In the illustrations in this book, we very frequently present only a set of points without any attempt to join them sequentially by a curve.)

What is a fractal spiral? I consulted three books on fractals. Though there were pictures, there was no definition.

How does one proceed from a common, perhaps visual experience, to a mathematical definition, or at least to a common but perhaps unspoken set of agreements? What is a straight line? What is a curve? What is a number? What is a set? What is a polyhedron? (If you want an amusing and deeply philosophical discussion of the historical attempts to define a polyhedron, read Imre Lakatos' masterpiece *Proofs and Refutations* [Lakatos 1976].) What is a quasi-crystal?[27] What is a spiral tiling?[28] What is probability? What is entropy? What is chaos?[29]

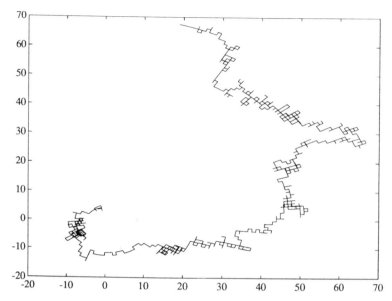

Figure 14: Discretely randomized Theodorus spiral.

$$d = \text{floor}(7 * \text{rand}); z = z + i^d z/|z|.$$

One makes normative definitions, that is, definitions that preserve some desirable property. For example, as emphasized by
Lakatos, in defining a polyhedron, one would like the Euler–
Descartes theorem on vertices, edges, and faces to survive the
definition. There must be a sense in which a definition, once
promulgated, proves fruitful and stabilizes. This process may
take centuries, millennia; and since a definition is necessarily a
limitation – a finitization – the process is always unsatisfactory
and never ending.[30]

Would you say that the equation of a spiral cannot, in rect-
angular coordinates, be algebraic?[31]

Would you say that the product of two spirals is a third
spiral?[32]

What transformations would you say preserve spirals? Such
transformations form a semigroup. Perhaps there's a clue. Surely, the affine groups preserve spirality.[33]

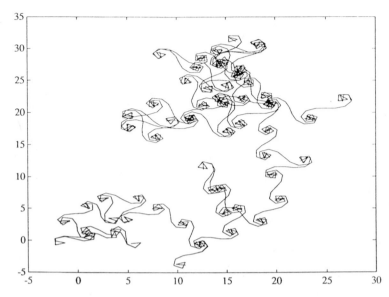

Figure 15: Trigonometric sums studied by Coutsias and Kazarinoff.

$$w = .3299; p = 1.781237; d = \exp(\pi i w j^p); z = z + d.$$

Would you say that if S is a spiral, its image under a diffeomorphism or a homeomorphism is also a spiral?

Would you say that the limit (in some sense) of a sequence of spirals is a spiral?

Would you say that if S is a spiral, and if you change a finite part of it in any way, it remains a spiral?

Would you say that if S is a spiral and you broke it in two like a cookie, the half would be a spiral?

Would you say that if you differentiate S, it remains a spiral? Would you say that if you take two (discrete) spirals and form their Kronecker product, a spiral will result?[34] (See figs. 10, 11a, 11b.) Note a spider spiral, integrated and differentiated?[35] (See fig. 9.)

It makes an interesting graphical exercise to take several elementary spirals and subject them to the functions that are

No, A st. line ≠ spiral

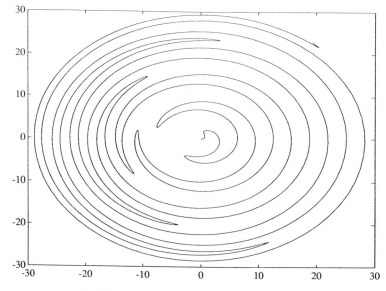

Figure 16: A labyrinth spiral.

commonly available in a computer package such as MATLAB. (See figs. 29a–31.)

Would you say that a curve is a spiral if it spins around indefinitely? Can a closed curve be a spiral?

How does one distinguish spirals from what some writers have called volutes, whorls, meanders, wanderings, doodles, noodles, tangles, or explosions? Why not introduce the term and concept of "spinner," following the example of E. Cesàro, who renamed Cornu's spiral the clothoid?[36]

Surely, you would not want a spiral to be a primitive undefined notion the way a point is. Then why all the wishy-washiness about its definition?

Listen to the words of Newton's teacher, Isaac Barrow[37]: Mathematicians "take up for contemplation those features of which they have in their minds clear and distinct ideas; they give these appropriate, adequate and unchanging names. . . . "

Brave words, spoken from the point of view of the "Monday-morning quarterback." Perhaps, as opposed to, say, the notion

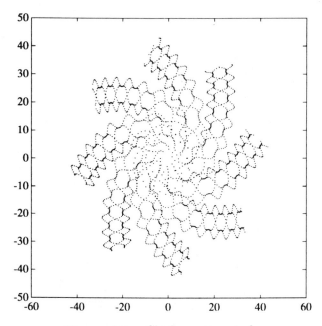

Figure 17: Chicken-wire spiral.

$$a = \exp(\pi i/4); \quad b = \sin(j) + \sin j/5; \quad z = az + bz/|z|.$$

of a group, which has been quite stable for almost two centuries, mathematicians do not yet have "clear and distinct ideas" about spirals. Perhaps it is not important; perhaps it would be counterproductive to introduce a definition.

But the story is much more complex. Perhaps part of the answer lies in an observation of Charles Sanders Peirce. Peirce once wrote that technical words should be introduced that are "...so unattractive that loose thinkers are not tempted to use them."[38]

Now "spiral" is an ancient word, an attractive word. It has warmth, juicyness, onomatopoeia; it suggests the lively playfulness of the universe. Everybody loves a spiral, wants to have a spiral of their own.[39] We refuse, therefore, in the terminology of philosopher W. V. O. Quine, to allow its meaning to be "regimented."[40]

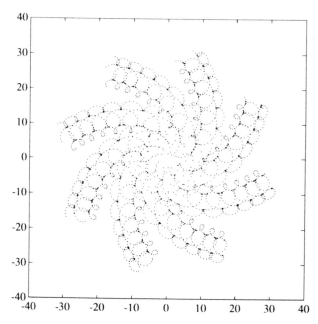

Figure 18:   Chicken-wire spiral. Note multiple loopings in each
spiral arm.

Of course, mathematics has ways of accommodating all tastes.
Mathematics slaps on its objects adjectives such as almost, sub,
semi, super, quasi, alpha-quasi, pseudo, true, faithful, fuzzy,
ultra, meta, strong, weak, degenerate, standard, generalized,
pre-, -like, -oid, incomplete, and hundreds more. And of course,
these adjectives can be piled up or multiplied, like so many
operators, so as to produce, for example, a weak, generalized,
fuzzy, alpha-quasi pre-spiral.[41]

But let me put some samples on display and ask you whether
or not you would call these figures spirals.[42] (See figs. 9–22.)

In the programs for the figures that follow, j stands for the
iteration number, and the programs are given in quasi-MATLAB
notation.

You may not see "spirals" in these figures[43] and in some
that follow. You may see instead comets, flowers, bugs, nebulae, pinwheels, meanders, explosions. Many of these figures

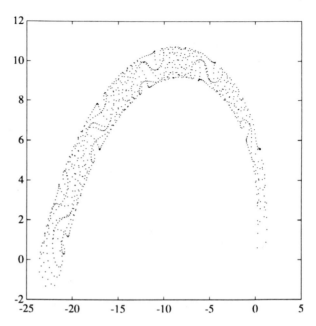

Figure 19:   Pyrotechnic spiral.

$$w = .7896;\, p = 1.1;\, d = \exp(\pi i w j^p);\, z = z + dz/|z|.$$

are instances of the "generalized spiral of Theodorus," and we
have therefore arrived at a neat theory that unifies bugs and
cosmic dust. But the closest I will claim to bona fide applications at the moment is to point out that the difference equation
for the spiral of Theodorus (see Lecture II) is identical to result of applying the Euler method for the numerical solution
of the differential equation $z' = z/|z|$, with a step size $h = 1$.
The spiral that emerges (instead of the true solution, which is
a circle) is a standard demonstration of the truncation errors of
the discretization process.[44]

There is another approach; a modern approach to deciding

what a spiral is. Build yourself a "spiral recognizer." Ask a
hundred people to draw pictures of spirals and pictures of nonspirals. Then write a program that allows the computer to dis-
criminate between the spirals and the nonspirals. This is easy

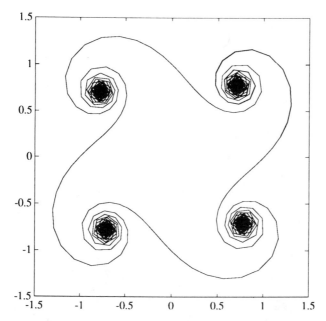

Figure 20:   Sums of quadratic exponentials inspired by the carving on an ancient monumental stone from Gottland, Sweden.

$r = .0234i; s = .0959(1 + i); w = w + \exp(rj^2); v = \exp(sw); \text{plot } v.$

to do: Just discretize the curve and consider it to be a point in hyperspace in a finite (and not too large) number of dimensions. Then go into the stacks of your science library or sit in front of your terminal and pray to the Shade of Archimedes that the two sets – the spirals and the nonspirals – can be separated by something simple such as a hyperplane. Or a hyper-something.

Well, that does not seem to be the way it works. How does it work? I want to teach a computer what a spiral is.

So I inquired of Dr. Kevin Manbeck, who has been working in my department on identification problems of arteriograms (of the heart). His work involves such things as dynamic programming techniques, and he answered me along the following lines.[45]

There would be two phases to the process. The learning phase and the discriminator phase. In the learning phase, we would

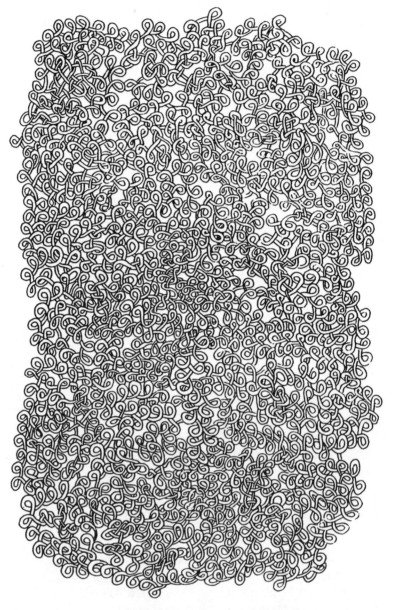

Figure 21:   Mendès-France's Tasmanian noodle (thickness added for aesthetic reasons). *Courtesy of M. Mendès-France.*

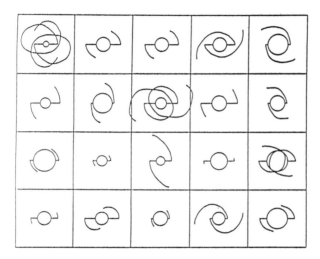

Figure 22: Spiral galaxy templates (prototypes). *From B. D. Ripley and A. I. Sutherland [1990].*

need first to work forward from a few templates of what you might call standard spirals.

Then we would need to specify what are the allowable transformations or perturbations. Having done this, we could move forward and create images we would call spirals. Then, in the discriminator phase, given a putative spiral, we could go backward and ask whether it conformed to some allowable template modulo the allowable perturbations.

I answered: "Thank you. But there are days when I think I have an infinity of spiral templates in my head. Moreover, I haven't been able to make much progress in specifying all the transformations I think might be allowable. There are even days," I added, "when I think that all curves are spirals."

"In that case, wouldn't you say that the problem of discrimination is trivial?"

Properly rebuffed, I retorted that I thought one of the major problems of the computerization of mathematics was to get a computer to recognize automatically what was mathematics and what wasn't.[46]

Lecture II

Lessons from Euler's Ghost

I return to what, if we are prepared to believe the fantasy of Herr P. 8
Anderhub, may very well be the first spiral discussed in a mathematical context: Professor Hlawka's Quadratwurzelschnecke. I p. 7
propose to call it the *spiral of Theodorus*. After all, if it is the
Ur-spiral, the granddaddy of mathematical spirals, it deserves
a classical name.

I had better begin with a word about Theodorus. Not too
much is known about this gentleman.[47] His approximate dates
are 465 to 399 B.C. (so he lived two centuries before Archimedes).
He was born in Cyrene, which was then a flourishing and substantial Greek colony just south of Greece on the North African
coast. Theodorus was the teacher of both Plato and Theaetetus.
He started out life as a philosopher and switched to mathematics.

I place the Theodorus spiral in the complex plane and define
its vertices z_n in iterative fashion:

$$z_{n+1} = z_n + iz_n/|z_n|, i = \sqrt{-1}, \tag{2.1}$$

where $z_0 = 1$, for example.

Now this is a first-order (system, if regarded in real coordinates) nonlinear difference equation.[48] But since we are completely aware that the spiral was used to create the square roots of the integers, we know that

$$|z_n| = \sqrt{n+1}. \tag{2.2}$$

We can therefore replace the original equation (2.1) with

$$z_{n+1} = (1 + i/\sqrt{n+1})z_n, n = 0, 1, 2, \ldots, 3, \tag{2.3}$$

and this is a linear, homogeneous difference equation, but with nonconstant coefficients. Recalling that the factorials $n!$ arise from a linear, nonconstant difference equation,

$$z_{n+1} = nz_n, \tag{2.4}$$

and that the gamma function lurks in the background of (2.4), we anticipate that interesting difficulties lie ahead with (2.1) or even (2.3).

We can write immediately from (2.3) a nice formula for z_n:

$$z_n = \prod_{k=1}^{n} (1 + i\sqrt{k}), \tag{2.5}$$

and on this basis, or on the basis of (2.3), we can study initially what I shall call the *discrete spiral of Theodorus* (Fig. 1). Equations (2.1) or (2.3) are also easy ways of computing it and getting a picture.

In Lecture I, we have already mentioned some facts about z_n. Now the problem suggests itself immediately, just as it did in 1729 to Christian Goldbach in his correspondence with Euler: How does one interpolate to the values z_n for noninteger values of n?[49] Stated differently: How can we draw a continuous curve through the sequence of points $z_0, z_1, z_2, \ldots$?

From the viewpoint of general interpolation theory, where's the problem? Draw the curve any way you like, and there's an end to the matter. Of course, there are an infinity of ways in which you can do it. One simply defines a policy or a strategy. The graphics facility of my MATLAB matrix package has its own

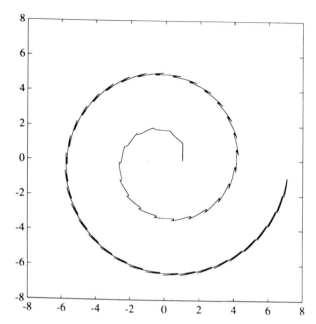

Figure 23:   Continuation of linear segment via the Theodorus difference equation.

default policy and will oblige by connecting up the points automatically with straight line segments. Of course the resulting curve will then have corners.

Oh, you want smoothness? Well, then, why don't we employ parametric cubic splines? (A cubic spline is a function that is a piecewise cubic and is of continuity class C^2.)[50]

But demands on the interpolant might still be multiplied. It might be required that the interpolant satisfy the difference equation (2.3) for noninteger values of n. It might be required that the interpolant be analytic or that both of these hold simultaneously.

If only the first condition is required, we may proceed as follows. Define a continuous curve from z_0 to z_1 completely arbitrarily, and use the difference equation to fill in the values between z_1 and z_2, z_2 and z_3, ...

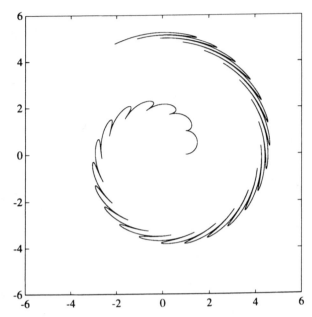

Figure 24: Continuation of semicircular segment via the
Theodorus difference equation.

The adjoining figures show what happens when you start out
with a straight line segment (which seems a very natural thing
to try), and with a semicircle (which seems unnatural). Note
the folding of the arc that occurs as the extrapolation via the
difference equation moves forward. This suggests a bit of study.

In fig. 23, a straight line is drawn through the first two points.
The figure is then filled in using the Theodorus difference equation (2.1).

In fig. 24, the first two Theodorus points are the endpoints of a
semicircle. The remaining figure is filled in using the Theodorus
difference equation (2.1).

We may also adopt a mixed strategy of interpolation: interpolate by means of a parametric cubic spline, passing through,
say, fifty points, and then use the portion between z_0 and z_1
to continue the curve forward by the difference equation. This
policy seems to yield good results as far as I have carried it.

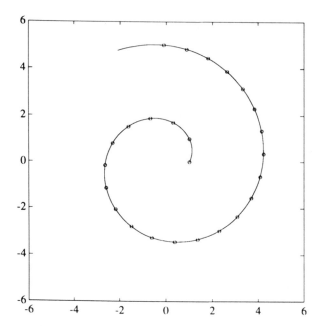

Figure 25:   Discrete spiral of Theodorus interpolated by cubic spline through six points and carried forward.

A parametric cubic spline has been passed through the first six Theodorus points. The first two points are thereby connected by a certain arc. The remainder of the figure is completed by using the difference equation on this arc. Note the smoothness and the fact that visually the curve is "spirally convex," whatever that term might mean mathematically.[51]

But the mathematical heart may require still more: an analytic solution in the sense of the "special function theory" of the eighteenth-century mathematicians. Those fellows started from finite pileups of factors and solved the interpolation to noninteger values by means of infinite pileups of factors, using a certain cancellation trick. Having recalled and studied Euler's infinite product for the gamma function

$$[(2/1)^n 1/(n+1)][(3/2)^n 2/(n+2)][(4/3)^n 3/(n+3)] \cdots = n!,$$
$$(2.6)$$

it occurred to me how to proceed with the difference equation
(2.3), and I wrote down

$$T(a) = \prod_{k=1}^{\infty} (1 + i/\sqrt{k})/(1 + i/\sqrt{k + a}). \qquad (2.7)$$

Note that the individual terms, $u_k(a)$, in (2.7) are, asymptotically,

$$u_k(a) = 1 + 0(k^{-3/2}), \qquad (2.8)$$

so that the product is absolutely (though very slowly) convergent.

The function $T(a)$ will be called the *Theodorus function*.[52]
(See fig. 26.) It is readily observed that it satisfies the difference
equation (2.3), it satisfies $|T(a)| = \sqrt{a + 1}$ for $a > -1$, it is an
analytic function of a and allows an analytic continuation of T
into the complex a-plane.

Setting $H(a) = T'(a)/T(a)$, one arrives at

$$H(a + 1) - H(a) = (-i/2)/((a + 1)^{3/2} + i(a + 1)^{1/2}), \quad (2.9)$$

which is a linear, constant-coefficient, nonhomogeneous difference equation. The function $H(a)$ may be considered analogous
to the ψ (or digamma) function in the theory of the Γ function.
In the case of the ψ function, the right-hand side is $1/a$ and
hence much simpler than (2.9).

I am proposing the Theodorus function (2.7) as *the* solution
to the difference equation.

Now we are in a sort of dilemma. We may add any function
of period 1 whatever to a solution of (2.9) and arrive at another solution. So there are an infinity of solutions; there are
even an infinity of analytic solutions.[53] More than this, there
are several disparate general theories of the solution of such
a difference equation. Among them, we may mention the theory of N. E. Nörlund [1924; 1929], the theory of E. Artin [1931]
(worked out for $\log \Gamma$ and greatly generalized by W. Krull [1948;
1949]). These authors speak of "principal solutions," "normal

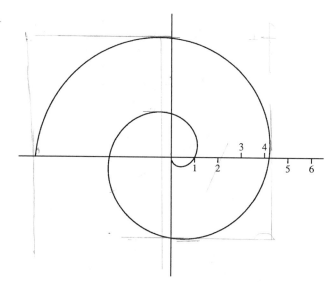

Figure 26:   An analytic spiral of Theodorus, as computed from the infinite product (2.7). *Courtesy of W. Gautschi.*

solutions," and so on. There are also uniqueness theorems along the line of Laugwitz and Rodewald [1987].[54]

Problem: In what way is the Theodorus function (2.7) distinguished from among the solutions of the difference equation? Certainly it is distinguished in terms of the simplicity of derivation, but what distinguishes it organically? Is it a solution that is "natural," or "nice," or, in the terminology of automotive design, "sweet"; the solution that our eye seems to "impose"?[55]

The analogous question for the gamma function was not answered until 1922. Now known as the Bohr–Mollerup–Artin theorem, it says that the Euler gamma function $\Gamma(a)$ is the unique function defined for $a > 0$ for which $\Gamma(1) = 1$, which satisfies $a\Gamma(a) = \Gamma(a + 1)$, and for which $\log \Gamma(a)$ is convex.[56]

It would be nice to have an analogous characterization of $T(a)$.

In possession, then, of a "brand new" special function $T(a)$, I wanted first of all to compute it, and, regarding its slope as it crosses the 0^0 ray as a fundamental world constant and calling this world constant T in honor of Theodorus, I wanted to determine T to about eight or ten figures to the right of the decimal point. A bit of paper work revealed:

$$T = \sum_{k=l}^{\infty} 1/(k^{3/2} + k^{1/2}).^{57} \qquad (2.10)$$

I wanted ultimately, if possible, to relate $T(a)$ to individual members of that large family of special functions whose properties have been worked out in depth and that, at one time, were certainly part of the working vocabulary of all analysts.

I tackled the computation naively. Computer power is so immense these days that one wonders whether, in a particular case at hand, it pays to waste time on being clever.[58]

The series and products above converge like $n^{-1/2}$, where n is the number of terms taken. So I turned on the switch and let the computer run to a million terms while I went out to lunch. This, of course, was not enough: I hardly got three significant figures. I applied a Richardson speedup procedure and got three additional figures. Wanting rather more figures (you will see why in the Notes) and being both impatient and lazy, I contacted George Phillips at the University of St. Andrews, who is an expert on how convergence can be speeded up. I knew that Phillips had shown how, if Archimedes had been really clever in his computations of π, he might have squeezed out fifteen more correct decimal digits out of the raw data that he derived from polygons of ninety-six sides. If Phillips could do that much for Archimedes, he certainly could do as much for Theodorus.[59]

Phillips was most obliging and, by return mail, sent me the following equation, which he obtained by using the Euler–Maclaurin summation formula[60]:

$$T = T_{n-1} + R_n, \qquad (2.11)$$

where T_{n-1} is the sum of the first $n-1$ terms of (2.10), while
the remainder R_n is approximated by

$$R_n = n^{-1/2}(2 - (1/6)n^{-1} + (1/40)n^{-2} + (1/168)n^{-3}$$
$$-(5/1152)n^{-4} - (3/1408)n^{-5}$$
$$-((303/66560)n^{-6}) + 0(n^{-7.5}). \tag{2.12}$$

This strategy is very efficient. With $n = 11$, it yields the value

$$T = 1.860025078 \tag{2.13},$$

which is a remarkably good result for so little computation.
(The last digit to the right of the point should be 9.)

9 x

Returning to my earlier approximative attempts, a mystery
emerges: If you take the discrete spiral of Theodorus and spline
it, say, through its first six points, you get a lovely curve that
fits the points. However, will or will not this spline coincide with
the "canonical" curve given by the infinite product (2.7)?

This question is nice because either way you answer, you have
a mystery: something to explain. If it does, why does it? If it
doesn't, why doesn't it?

As mentioned, I put a spline through the first six points of
Theodorus. I got a very nice curve that appears spiral-convex.
But its slope at the beginning ($a = 0$) is $2.0729\ldots$, whereas
the slope of the infinite product (2.7) is $1.860025\ldots$. End of
the matter. (See fig. 25.)

Could we do better with splines with more points or splines
of a higher degree? But recall: Cubic splines that simply fit a
bunch of points still have two degrees of freedom left. Perhaps
the "cantilevered" spline in which the initial and final slopes
are specified is what one should work with.

Computation is one thing, and the identification of $T(a)$ is
another matter, and it still eluded me. The Spirit of Euler in-
fused me constantly, but contributed nothing toward the solution. The mistake I made was that I had been consulting
the wrong Swiss mathematician. I should have consulted the

Swiss-born-and-trained American mathematician, Walter Gaut-schi, who walked into my office one day before giving a lecture at our colloquium. I showed him the series for T, and within the week, he had computed T to twenty places in what I consider to be an absolute gem of numerical analysis[61]; and perhaps more importantly, in the course of this work, he had also identified $T(a)$.[62] This development is presented in Supplement A. Here are a few highlights.

Note that

$$1/(s^{1/2} + s^{3/2}) = (1/s^{1/2})(1/(1+s)) \qquad (2.14)$$

and therefore the left-hand side of this equation equals $\mathcal{L}(\frac{1}{\sqrt{\pi t}} *e^{-t})$, where $\mathcal{L}$ designates the Laplace transform and $*$ the op-eration of convolution.

From this beginning observation, Gautchi was able to express T, and similarly for $T(a)$, as

$$T = \frac{2}{\sqrt{\pi}} \int_0^\infty t^{1/2} e(t) g(t) dt, \qquad (2.15)$$

where $e(t)$ is the Einstein function

$$e(t) = t/(e^t - 1), \qquad (2.16)$$

$$g(t) = \frac{1}{\sqrt{t}} F(\sqrt{t}), \qquad (2.17)$$

and where F is Dawson's integral[63]

$$F(t) = \exp(t^{-2}) \int_0^t \exp(t^2) dt. \qquad (2.18)$$

Now, to compute T, Gautschi suggests that one use Gaussian rules of approximate integration with either the weight $t^{1/2} e(t)$ or the weight $t^{1/2} \exp(-t)$. The former is a weight function for which the recurrence coefficients of the related orthogonal poly-nomials are not expressible in closed form; hence a special com-putational strategy must be invoked – one that has been worked out in great detail by Gautschi (see Supplement A).[64]

Gautschi's basic identity may be written symbolically so as to display a lineup of famous names: the differentiated argument of

$$\text{Theodorus} = \text{Laplace (Einstein} \cdot \text{Dawson)}.^{65}$$

The "identification" of $T(a)$ in terms of well-studied special functions opens up many questions such as: What is its character as an analytic function of a complex variable a? What other (if any) functional equations does it satisfy? Is the constant T transcendental? Is $T(a)$ transcendentally transcendental, as one might suppose, thinking that (2.3) is more complicated than (2.4)?[66]

Allowing one's imagination to wander freely, since there are connections in this part of special function theory to the Riemann zeta function, wouldn't it be grand to be able to come up with some fact about the spiral of Theodorus whose truth depends upon the assumption of the Riemann hypothesis!

Before leaving the topic of continuous (or fractional) interpolation to the orbit of a difference equation, one should mention a certain generalization of the spiral of Bernoulli that, for short, might be very well called the *spiral of Bernoulli–Schroeder*. This relates to the iteration of functions of a complex variable z that are analytic in a neighborhood of $z = 0$.

Suppose that $f(z) = az + \cdots$ is analytic in a neighborhood of $z = 0$, and suppose further that we can find a function $H(z) = z + \cdots$, also analytic in a neighborhood of $z = 0$, and for which

$$H(f(z)) = aH(z). \tag{2.19}$$

Again, in a neighborhood of $z = 0$, this leads to

$$f(z) = H^{-1}(aH(z)). \tag{2.20}$$

By analogy to matrices, we can say that H "diagonalizes" the function f or that f is conjugate to the linear function $L(z) = az$ under H. The function H will be called the *Schroeder function* for f. If now, we have a sequence of points $z_0, z_1, \ldots$ defined by

$$z_{n+1} = f(z_n); z_0 = \text{starting point}; \qquad (2.21)$$

then, formally at least,

$$z_n = H^{-1}(a^n H(z_0)), n = 0, 1, 2, \ldots. \qquad (2.22)$$

If one now sets $a = r \exp(i\theta)$,

$$z(t) = H^{-1}(r^t \exp(it\theta)H(z_0)), -\infty < t < \infty \qquad (2.23)$$

gives us a continuous interpolation to the discrete orbit and which satisfies the difference equation $z(t + 1) = f(z(t))$ for all t.

If $0 < r < 1$, then $z(t)$ will be called the *spiral of Bernoulli–Schroeder* corresponding to the iteration function f and the initial value z_0.

When $r \neq 1$, the Schroeder function exists and is constructible by iteration. When $r = 1$, the theory of the Schroeder function is very deep, leading to the problem of small divisors and the famous theorem of Siegel–Moser. It is known that for almost all a on the unit circle, a Schroeder function exists.[67]

Lecture III

Theodorus Goes Wild[68]

"Man muss immer generalisieren," wrote C. G. J. Jacobi. Mathematicians should always generalize, and moved by this directive and without excessive exertion of the imagination, one writes down

$$z_{n+1} = az_n + bz_n/|z_n|, \qquad (3.1)$$

for a and b arbitrary complex numbers, and hopes that this yields something interesting. It does. The dynamics of (3.1) are nontrivial.

One may even move out of the space of one complex variable into a vector formulation and write down

$$v_{n+1} = Av_n + Bv_n/ \parallel v_n \parallel +c, \qquad (3.2)$$

where A and B are square matrices (with real or complex elements), c a column vector, and v_n a sequence of column vectors of appropriate dimension, and the norm equals some accessible and interesting vector norm. A possible interpretation of

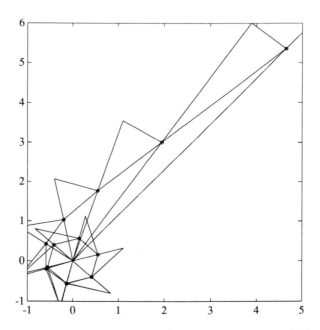

Figure 27: Pinwheel spiral (construction lines added).

(3.2), with $c = 0$ (or of any of the previous special cases), in the spirit of mathematical modeling was pointed out to me by Neil Miller: v_n is the current state of a number of interacting populations (say, age-classified populations). A is the geometric interpopulation growth rate matrix. Since $v_n / \| v_n \|$ has norm 1, the term $B v_n / \| v_n \|$ can be regarded as an migration term of bounded size and distributed linearly according to "proportion" of the individual strengths of the current populations. The matrix A might be assumed, for example, to be a Leslie matrix (all nonzero elements in the first row and first subdiagonal) and the 1-norm used.[69]

The full panoply of matrix theory is now available to suggest problems, and MATLAB, an extremely friendly matrix package, is available for numerical experiments.

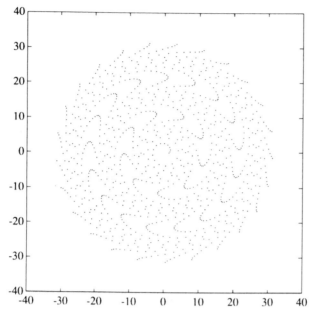

Figure 28:   The marigold spiral.
$$z_{n+1} = az_n + bz_n/|z_n|$$
$$a = \exp(\pi i/4) = (1/\sqrt{2})(1+i)$$
$$b = \bar{a}.$$

One could even contemplate a formulation that might occasionally be more convenient:

$$Gv_{n+1} = Av_nB + Cv_nD/\parallel v_n \parallel, \qquad (3.3)$$

where all the symbols now represent square matrices[70]; or the form

$$v_{n+1} = Av_n + p_nBv_n, \qquad (3.4)$$

where p_n is a sequence of scalars. And, of course, one can also study stochastic versions of these equations.[71]

However, one must keep the generalization process under control, for if one generalizes just a bit too much, one can, in theory, generate all possible sequences of vectors from such a recurrence.[72]

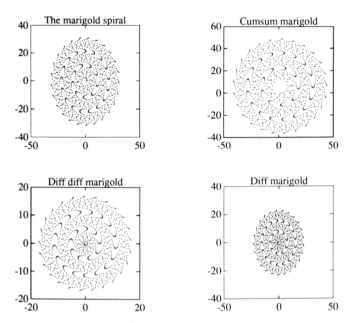

Figure 29a: Integrated and differentiated marigold spirals.

Returning to the "complex variable case" (3.1), notice that it contains within itself the discrete spiral of Theodorus, ($a = 1; b = i$), the discrete spiral of Bernoulli, $|a| \neq 1$ and $\text{Im}(a) \neq 0; b = 0$, and the discrete spiral of Archimedes, ($|a| = 1, a \neq 1, -1; b = sa, s > 0$). This a nice unification.

Thus, for example, the "pinwheel spiral" emerges from the selection $a = .5, b = .5i$. (See fig. 27.)

The construction lines suggest how the iterates can be created with a ruler and a right angle.

Some of the iterations generate figures that are quite striking, visually speaking, particularly when the discrete values are plotted as discrete dots and are not connected up by line segments. The eye then "connects" up the dots in its own way, often organizing the total figure into many spirals, etc. (See fig. 32.) This phenomenon is related to what in signal processing is called the *aliasing* that results from discrete sampling.[73]

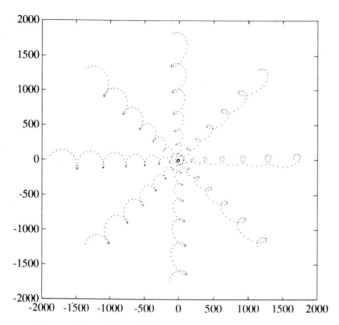

Figure 29b: $5T + M + 3TM$; M = marigold, T = Theodorus.

With $a = \exp(\pi i/4)$, $b = \bar{a}$, one obtains a mathematical marigold, (why shouldn't we call it the *marigold of Theodorus?* with a beautiful inner texture. (See fig. 28.)

The successive rings of petals are particularly nice. Now that you know what the iteration produces, could you have predicted it from the equation? What can be said theoretically about the rings of petals?[74] Is it really important to prove anything about them theoretically?

It is interesting to consider a fringe area where the eye/brain is confronted with the dilemma of whether to organize the material into one "traditional" spiral or into multiple "fan blade" spirals. (See fig. 31.) My experience is that upon steady observation, the organization shifts from one to the other and back again, and constitutes yet another instance of a visual paradox or illusion.

Notice also the recurrence

$$|z_{n+1}|^2 = |a|^2|z_n|^2 + 2\,\mathrm{re}(\bar{a}b)|z_n| + |b|^2. \qquad (3.5)$$

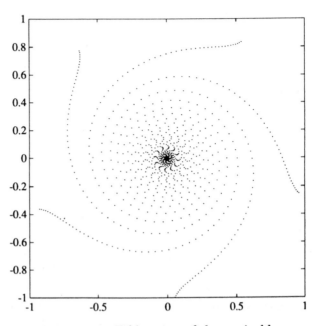

Figure 30:   Fifth power of the marigold.

This means that the dynamics of $|z_n|$ follows a one dimensional real iteration (the square root of a quadratic). This carries with it the possibility that orbits (shall I call them *spirals?*) possess fixed points, cyclic points, period doubling, invariant curves and measures, bifurcation, strange attractors, basins, equilibrium distributions, and so forth; in short, the whole panoply of features associated with chaos theory that have received intense study in the past twenty or so years.[75]

The values $a = 1.3192 + .4751i$; $b = -a$; $z_0 = 1.5$ generate Figure 33. Chaotic? Yes, and yet the eye discerns patterns. The eye "wants" to see. It would be an interesting exercise to reduce what it sees in this figure to a purely verbal description; two rings surely and with increasing density as one approaches the center.

So chaos lurks just a short distance from Theodorus. I shall leave it to others to discuss this question properly; for example, to Jeffery Leader, whose doctoral thesis deals with some aspects

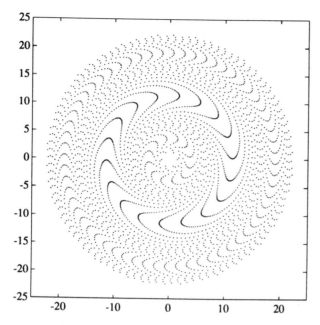

Figure 31: Marigold symmetrized in kaleidoscopic fashion.

of this question (particularly the case $A = 0$ in (3.6), which is related to the power method for eigenvalue computation); to Arieh Iserles, who has undertaken a profound study of these iterations (see Supplement B); and, of course, to whoever wants to pitch in.

It should be remarked that the eye discerns many aspects to the patterns that are generated by recurrence; for example, there are aspects that have to do with the "texture" of the figure. Until recently, such aspects have been almost totally ignored in favor of those qualities that arise and are of proved historical importance in classical dynamics.[76] But presumably, the textural aspects are also open to discussion, to definition, and to theoremization. A full classification of the orbits of (3.2) should take into consideration all the "natural" distinctions that the eye makes, and this would be very difficult indeed.[77]

A note on figures in dimension > 2: Employing standard techniques of computer graphics, one may exhibit projections of

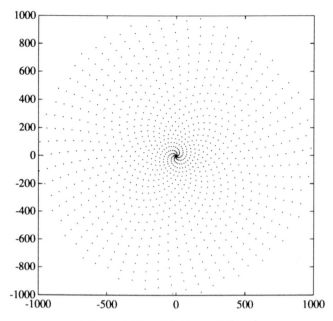

Figure 32: Illusion spiral.

$$z_{n+1} = az_n + bz_n/|z_n|; \quad a = .6 + .8i, \quad b = .65 + .7599i.$$

higher-dimensional sequences and animate them, making movies, for example, by rotating the projections about selected axes. The results are often spectacular. One may also render them by applying "skin" in a variety of ways, applying color, and employing reflectivity strategies. It would be an interesting problem to ask for analytic interpolants for these higher-dimensional objects.[78]

It should be further remarked that in "most" instances, but with some interesting exceptions, the root-quadratic iteration (3.1) does not possess a simple analytic solution ($\sqrt{n+1}$ in the Theodorus case), and so the passage from a nonlinear system to a linear system, and from a discrete orbit to an analytic orbit, cannot be worked out and linked to special function theory along the lines of Euler–Gautschi.

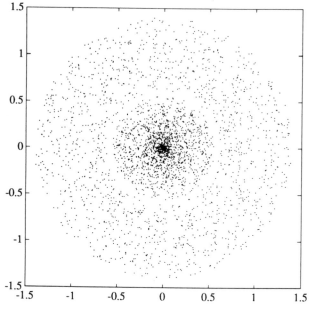

Figure 33:   Chaos (?) spiral.

We now take a closer look at the iteration

$$v_{n+1} = Av_n + Bv_n/ \parallel v_n \parallel . \qquad (3.6)$$

First, some figures. Figures 34–39 derive from the work of
Jeffrey Leader. Figures 40–46 derive from the work of Arieh
Iserles.

As far as I am aware, a complete analysis of this iteration has
not yet been made.

Some general observations. It is important to eliminate as
many parameters as possible. Call $(A, B; v_n)$ a *Theodorus triple,*
meaning matrices and vectors that are linked by the basic difference equation above. Let σ be a complex scalar and Ω a unitary matrix. Designate the conjugate transpose by *. Then if
$(A, B; v_n)$ is a Theodorus triple, so is $(\Omega A \Omega^*, |\sigma| \Omega B \Omega^*; \sigma \Omega v_n)$.
We could factor out the group of rotations and scalings this
way.[79]

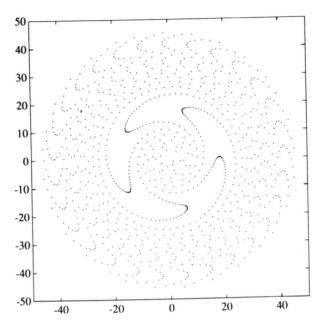

Figure 34:   Penta-fanblade spiral.
$r = .2347$, $s = .9721$, $t = 1.3747$, $u = -.3319$,
$A = [r, s; -s, r]$, $B = [t, u; -u, t]$.

$B = 0$. This is the linear case and is completely worked out theoretically. But even in this case, there may be special problems that suggest themselves and could be attacked. For example, the numerical handling of Jordan blocks is highly unstable and notoriously difficult. (MATLAB will not Jordanize a nondiagonalizable matrix.) The Book of Computation (including even the chapter on the simple arithmetic operations $+$, $-$, $\cdot$, $/$) seems never to be closed. New generations of computers (e.g., vectorization, VLSI) create new challenges and occasionally disinter old possibilities.

One should also remark, en passant, that if $\det(A) \neq 0$, the problem of analytic interpolation to the orbit v_n has a natural solution as follows. For all square matrices M, $\exp(M)$ exists,

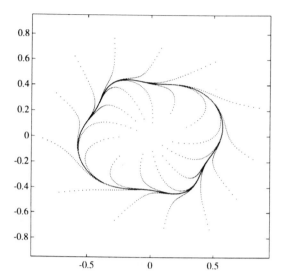

Figure 35:   Exhibiting invariant curve.
$A = [.91, .71; -.65, .58]; \; B = -A.$

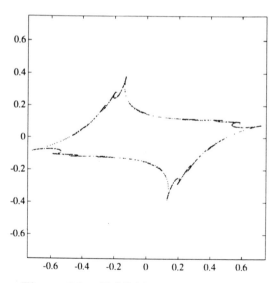

Figure 36:   Exhibiting invariant curve.
$A = [.2128, .1304; .7147, .0910]; \; B = \text{transpose}\,(-A).$

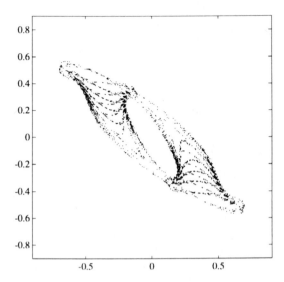

Figure 37: The double cornucopia spiral.
$A = [.51, -.1; .08, -.37]; \ B = [-.12, .71; -.22, -.45].$

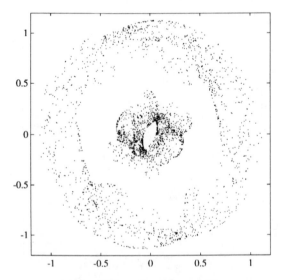

Figure 38: Double wreath.
$A = [1, -1; 1, 1]; \ B = [-.9, .8; -.8, -1].$

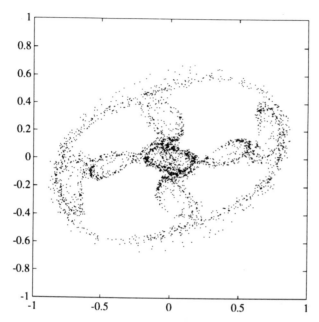

Figure 39:   Steering wheel.
$A = [-.1473, .6316; .8847, .2727]; \quad B = \text{transpose}(-A)$.

and if M and N commute, then $\exp(M)\exp(N) = \exp(M+N)$. If $\det(A) \neq 0$, there exist matrices – call them $\log(A)$ – such that $\exp(\log(A)) = A$.[80] Select such a $\log(A)$. For $-\infty < t < \infty$, a fractional power of A – call it A^t – can now be defined by $A^t = \exp(t\log(A))$ and satisfies the law of exponents $A^t A^u = A^{t+u}$. Moreover, if t is an integer, then A^t coincides with the usual definition of a matrix power. Since in the case under discussion ($B = 0$), we have $v_n = A^n v_0$, then $v_t = A^t v_0$ provides an analytic interpolation that satisfies the difference equation $v_{t+1} = Av_t$, $-\infty < t < \infty$.

If $\det(A) = 0$, problems may arise with the interpolation problem as posed. For example, if A is nilpotent, (i.e., there exists an integer $m \geq 2$ such that $A^m = 0$ and $A^{m-1} \neq 0$), then $v_n = 0$ for $n > m$, while in general, $v_n \neq 0$ for $n \leq m$.

$A = 0$. This is related to the power method for the eigenvalues/vectors of B. If B is nonsingular, the orbit is the

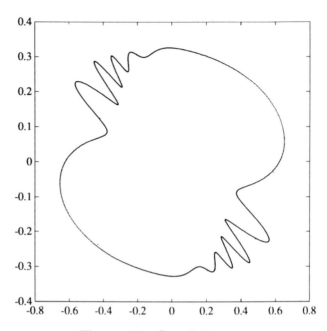

Figure 40: Invariant curve.
$A = sP, \ s = .7, \ P = [1.5, .71; -.41, .58], \ B = -A.$

(hyper)ellipse $x^* M x = 1$, where $M = \text{inv}(BB^*)^{-1}$. The lengths of the semiaxes of the hyperellipse are precisely the singular values of B. The case of B singular has been studied in some detail by Leader.

The nonconvergent case (where B has several dominant complex eigenvalues of equal modulus that are not roots of unity or multiples) looks interesting chaoswise and could use some additional study.

Some simple but interesting things can be had on the cheap by working with vector/matrix norms, $\| \cdot \|$.

Theorem. *If $\| A \| < 1$, then any orbit is bounded. If $\| A \| \geq 1$, then an orbit may be bounded or unbounded.*

Theorem. *If the spectral radius of A is greater than 1, there is a vector v_0 for which the orbit is unbounded. However, it may happen that $\| A \| > 1$, and there is a bounded orbit.[81]*

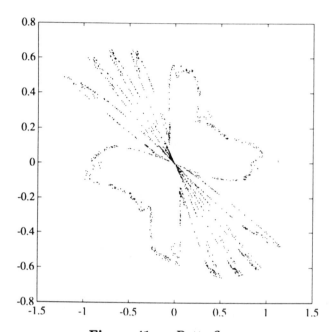

Figure 41a: Butterfly.
$A = sP, \; s = .95, \; P = [1.5, .71; -.41, .58]; \; B = -A.$

Theorem.   *If* $\|A\| \leq 1$, *then* $\|v_n\| \leq cn + d$.

Thus, the rate of growth is at most Archimedean.

Theorem.   *If B is nonsingular, then*

$$\|(v_{n+1} - Av_n)\| \geq 1/\|B^{-1}\|.$$

Hence, we cannot have $\lim v_n = 0$. *If v_∞ is a fixed point, then*

$$1/(\| \, I - A \, \|\| \, B^{-1} \, \|) \leq \| \, v_\infty \, \| \leq \| \, B \, \|\| \, (I - A)^{-1} \, \| \, .$$

Hence, if $A = I$ and B is nonsingular, no fixed point is possible.

What are necessary and sufficient conditions that the iteration be bounded? As far as I am aware, this question is still open.

What are necessary and sufficient conditions that the orbit

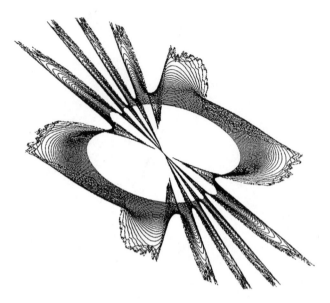

Figure 41b: Multi-butterfly.

converges to a (finite) fixed point? If the iteration has a fixed
point, then that point is a generalized eigenvector of the pair
$(I-A), B$ with a positive eigenvalue. If, for example, $B^{-1}(I-A)$
has no positive eigenvalue, then v_n cannot converge.

The following "pencil theorem" might be brought into play to
produce a nonexistence theorem for fixed points: Let A be positive definite symmetric. Let C be positive semidefinite symmetric. Then the roots of $\det(\gamma A - C) = 0$ are real and nonnegative
(see [Lancaster 1965, p. 100]).

If B is rank-deficient, there may be a finite number, an infinite
number, or no generalized eigenvalues. This case therefore leads
to a variety of possibilities.

The case A and B unitary deserves special treatment. If A
and B are unitary, $(AA^* = I; BB^* = I;$ and * indicates the
conjugate transpose), and if $A^*B + B^*A = \mu I, \mu =$ scalar, then,
as is easily shown, there is a (one-dimensional) iteration for
$\| V_n \|$. The case $\mu = 0, A$ and B unitary, A^*B skew-Hermitian

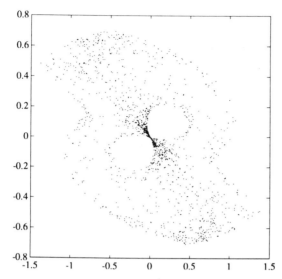

Figure 42:   Palette.

$A = sP, \ s = 1, \ P = [1.5, .71; -.41, .58]; \ B = -A.$

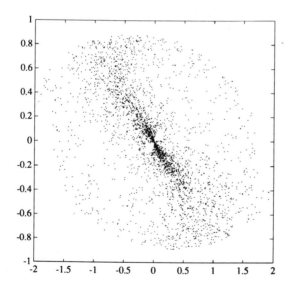

Figure 43:   Spooky palette.

$A = sP, \ s = 1.25, \ P = [1.5, .71; -.41, .58]; \ B = -A.$

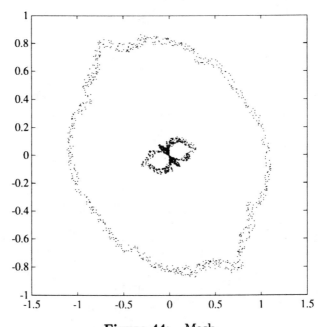

Figure 44: Mask.
$$A = sQ, \quad s = 1, \quad Q = [1.1, .71; -.65, .58]; \quad B = -A.$$

is already interesting and includes the selection

$$A = I, B = Y = \begin{pmatrix} 0 & 0 & 0 & 1 \\ 0 & 0 & -1 & 0 \\ 0 & 1 & 0 & 0 \\ -1 & 0 & 0 & 0 \end{pmatrix},$$

which may be regarded as the "two complex variable" Theodorus. If $\| v_0 \| = 1$, then $\| v_n \| = \sqrt{n+1}$, so that the system can be reduced to one that is linear but with nonconstant coefficients.

In the case under discussion, Iserles makes a full analysis of the iteration and shows, among other things, how, in even dimensions, to select A and B so that the normalized iterants will be equidistributed on the direct product of two circles, and hence cannot be equidistributed on the surface of the 4-sphere.

If A and B are unitary and AB^* is skew and A and B commute (they need not commute: Just take B as a unitary

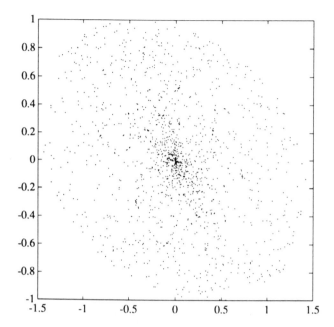

Figure 45:   Embossed shield.
$A = sQ, \ s = 1.15, \ Q = [1.1, .71; -.65, .58]; \ B = -A.$

matrix at random and $A = YB$), we may write an analytic
interpolation formula for the multidimensional Theodorus in a
form analogous to (2.7):

$$v(a) = \prod_{k=1}^{k=\infty} \left(A + \frac{1}{\sqrt{k}}B\right)\left(A + \frac{1}{\sqrt{k+a}}B\right)^{-1}.$$

Interesting special cases include the n-dimensional (complex)
marigold where $A = wF, B = A^*; w^4 = -1, F = $ the (complex)
Fourier matrix of order n; that is, the matrix that performs the
discrete Fourier transform. If $\| v_0 \| = 1$, then, again, $\| v_n \| = \sqrt{n}$.[82]

The Case $B = \alpha A$
Among Iserles' many results in Supplement B, my favorite (a
hard choice!) relates to the existence of limit cycles in the case
$B = \alpha A$.

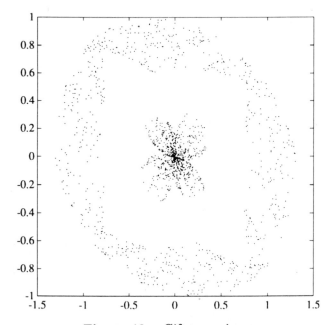

Figure 46:   Gift wrapping.
$A = [1.0465, .8165; -.7475, .6670]; \; B = -A.$

Theorem. *In R^2, if $B = \alpha A, \alpha > 0$, and if the spectral radius $p(A) \geq 1$, the sequence v_n diverges. If $p(A) < 1$, the sequence lies asymptotically on a limit cycle.*

Theorem. *If $B = -\alpha A$, and if $p(A)$, the spectral radius of A, is sufficiently small, then asymptotically and up to a linear transformation, the orbit lies on an invariant curve Γ defined by an equation of the form*

$$\Gamma(t) = \sum_{k=-\infty}^{k=\infty} g_k \exp(itk\theta)/(1 + p^{-1}\exp(itk\theta)), \quad -\pi \leq t \leq \pi,$$

and the "Fourier" coefficients g_k can be given an explicit representation in terms of certain hypergeometric functions.

Part II
Technical Developments

Supplement A

The Spiral of Theodorus, Special Functions, and Numerical Analysis

Walter Gautschi

A1. The Theodorus Function

The "Quadratwurzelschnecke," as smoothed out by P. J. Davis and represented in parametric form in the complex plane, has the equation

$$z = T(\alpha), \quad T(\alpha) = \prod_{k=1}^{\infty} \frac{1 + \frac{i}{\sqrt{k}}}{1 + \frac{i}{\sqrt{k+\alpha}}}, \quad \alpha > -1 \qquad (A1.1)$$

(cf. Eq. (2.7)). It spirals inward into the origin as α decreases from 0 to –1, and outward as α increases over positive values (cf. fig. 26). Replacing α in (A1.1) by $\alpha + 1$, and compensating for it in the denominator product by shifting the index k down by 1, immediately yields the useful recurrence relation (cf. (2.3))

$$T(\alpha + 1) = \left(1 + \frac{i}{\sqrt{\alpha + 1}}\right) T(\alpha), \quad \alpha > -1. \qquad (A1.2)$$

The function $T(\alpha)$ itself can be expressed in terms of infinite series by taking logarithms in (A1.1) and calculating the logarithmic derivative. This gives

$$\frac{T'(\alpha)}{T(\alpha)} = \frac{i}{2} \sum_{k=1}^{\infty} \frac{1}{(k+\alpha)^{3/2} + i(k+\alpha)}$$

$$= \frac{1}{2} \sum_{k=1}^{\infty} \frac{1}{(k+\alpha)(k+\alpha+1)} + \frac{i}{2} \sum_{k=1}^{\infty} \frac{1}{(k+\alpha)^{3/2} + (k+\alpha)^{1/2}}.$$

The first series is easily summed, its nth partial sum being

$$\sum_{k=1}^{n} \left(\frac{1}{k+\alpha} - \frac{1}{k+\alpha+1} \right) = \frac{1}{1+\alpha} - \frac{1}{n+\alpha+1}$$

and having the limit $(\alpha+1)^{-1}$ as $n \to \infty$. Thus,

$$\frac{T'(\alpha)}{T(\alpha)} = \frac{1}{2} \frac{1}{1+\alpha} + \frac{i}{2} \sum_{k=1}^{\infty} \frac{1}{(k+\alpha)^{3/2} + (k+\alpha)^{1/2}}.$$

Integration from 0 to α, noting that $T(0) = 1$, then yields

$$T(\alpha) = \sqrt{1+\alpha} \exp\left(\frac{i}{2} \int_0^{\alpha} U(\alpha) d\alpha \right), \qquad \text{(A1.3)}$$

where

$$U(\alpha) = \sum_{k=1}^{\infty} \frac{1}{(k+\alpha)^{3/2} + (k+\alpha)^{1/2}}. \qquad \text{(A1.4)}$$

In polar coordinates (r, φ), therefore, the spiral has the parametric representation

$$\begin{aligned} r &= \sqrt{1+\alpha}, \\ \varphi &= \frac{1}{2} \int_0^{\alpha} U(\alpha) d\alpha, \end{aligned} \qquad \alpha > -1.$$

Straightforward calculus can now be used to determine the more important geometric characteristics of the spiral, such as tangent vector, arc length, curvature, and so on. Inevitably, the function $U(\alpha)$ in (A1.4) and its derivative

$$U'(\alpha) = -\frac{1}{2}\sum_{k=1}^{\infty}(k+\alpha)^{-1/2}\frac{3(k+\alpha)+1}{(k+\alpha)(k+\alpha+1)^2} \qquad (A1.5)$$

(in quantities of second order) will figure prominently in the resulting expressions. For example, the line element turns out to be

$$ds = \frac{1}{2}\left(\frac{1+(1+\alpha)^2U^2(\alpha)}{1+\alpha}\right)^{1/2}d\alpha, \qquad (A1.6)$$

and the slope of the tangent vector to the spiral at $\alpha = 0$ (where it crosses the positive axis for the first time) is given by

$$U(0) = \sum_{k=1}^{\infty}\frac{1}{k^{3/2}+k^{1/2}}, \qquad (A1.7)$$

the "Theodorus constant" in Davis's terminology.

The infinite series in (A1.4), (A1.5) and (A1.7) all have one thing in common: they converge painfully slowly! To illustrate, take the series in (A1.7) defining the Theodorus constant and compute its first million partial sums. Here is what you will find:

n	nth partial sum
10	1.2615 . . .
100	1.6611 . . .
1000	1.7968 . . .
10000	1.8400 . . .
100000	1.8537 . . .
1000000	1.8580 . . .

With luck, one gets $U(0) \approx 1.86$ to three decimal digits! But what if ten correct decimals, or twenty, were wanted? Straight summation clearly would be hopeless. In the following sections, we develop a technique that is capable of computing such series to very high accuracy. (We will limit ourselves to twenty decimals.)

As far as computing the spiral itself is concerned (and also $U(\alpha)$ and $U'(\alpha)$, for that matter), it suffices to restrict attention to the interval $0 \leq \alpha < 1$, since for the remaining values of α, one can apply the recurrence relation (A1.2) in forward direction for the outward, and in backward direction for the inward, spiral. The arc $\{T(\alpha) : 0 \leq \alpha < 1\}$ is thus seen to be the core of the spiral – its heart, as it were – and it is also the most difficult piece to compute.

A2. A Class of Slowly Convergent Series and Their Summation by Integration

Factoring out $(k+\alpha)^{-1/2}$ in the general term of the series (A1.4) produces the simple rational function $(k + \alpha + 1)^{-1}$ as second factor. A slightly more complicated rational factor is exhibited in the series for U' in (A1.5). Both these series suggest considering a more general class of series, namely

$$S = \sum_{k=1}^{\infty} k^{\nu-1} r(k), \quad 0 < \nu < 1, \qquad (A2.1)$$

where $r(\cdot)$ is a rational function,

$$r(s) = \frac{p(s)}{q(s)}, \quad \deg p < \deg q. \qquad (A2.2)$$

The shift $k + \alpha$ characteristic in the series of Section A1 has been ignored in (A2.1), but will be incorporated a little bit later (cf. (A3.11)). Since shifting the index by a positive quantity is easily accomplished, we may assume without compromising generality that the zeros of the denominator polynomial q in (A2.2) all have nonpositive real parts:

$$\text{if } q(-a) = 0 \text{ then } \operatorname{Re} a \geq 0; \qquad (A2.3)$$

if this were not the case, we could in a preliminary step sum a few of the initial terms of the series (A2.1) directly and thereby achieve (A2.3) for the remaining series.

Having generalized the problem considerably, we now proceed to simplify it again! Any rational function of the form (A2.2), with real polynomials p and q, can be decomposed into partial fractions,

$$
r(s) = \sum_{\rho} \sum_{m=1}^{m_\rho} c_{\rho m}(s + a_\rho)^{-m}
$$
$$
+ \sum_{\gamma} \sum_{m=1}^{m_\gamma} [c_{\gamma m}(s + a_\gamma)^{-m} + \bar{c}_{\gamma m}(s + \bar{a}_\gamma)^{-m}]. \qquad (A2.4)
$$

Here the first sum extends over all real zeros $(-a_\rho)$ of q (having multiplicities m_ρ), while the second sum is over all pairs of conjugate complex zeros $(-a_\gamma, -\bar{a}_\gamma)$ (having multiplicities m_γ). The coefficients are complex, in general, except for those in the first sum, which are real. We assume that the decomposition (A2.4) has already been obtained; for relevant constructive methods, see [Henrici 1984, §7.1]. It is clear, then, that we can assume $r(\cdot)$ to have the form

$$
r(s) = \frac{1}{(s + a)^m}, \quad \operatorname{Re} a \geq 0, \quad \operatorname{Im} a \geq 0, \ m \geq 1, \qquad (A2.5)
$$

so that S in (A2.1) becomes

$$
S = \sum_{k=1}^{\infty} \frac{k^{\nu-1}}{(k + a)^m}. \qquad (A2.6)
$$

Our objective now is to transform the series (A2.6) into an integral. There are many ways this can be done; we choose to use the Laplace transform

$$
(\mathcal{L}f)(s) = \int_0^\infty e^{-st} f(t)dt. \qquad (A2.7)
$$

Suppose we can represent the general term in (A2.6), say a_k, as the Laplace transform of some function f, evaluated at the integer k,

$$a_k = (\mathcal{L}f)(k). \tag{A2.8}$$

Then

$$\sum_{k=1}^{\infty} a_k = \sum_{k=1}^{\infty} \int_0^{\infty} e^{-kt} f(t) dt$$

$$= \int_0^{\infty} f(t) e^{-t} \sum_{k=1}^{\infty} e^{-(k-1)t} dt$$

$$= \int_0^{\infty} \frac{f(t)}{t} \cdot \frac{t}{e^t - 1} dt,$$

assuming that the integral makes sense. Thus, letting

$$\epsilon(t) = \frac{t}{e^t - 1}, \tag{A2.9}$$

we have

$$\sum_{k=1}^{\infty} a_k = \int_0^{\infty} \epsilon(t) \cdot \frac{f(t)}{t} dt, \quad a_k = (\mathcal{L}f)(k). \tag{A2.10}$$

This is the desired integral representation.

Integrals of the type (A2.10), (A2.9) occur frequently in problems of solid-state physics, where they represent quantities of physical interest, such as the total energy of thermal vibrations of crystal lattices. In this context, the function ϵ in (A2.9) is known as Einstein's function. (Of course, ϵ is also the generating function of the Bernoulli numbers.) It was these physical applications that led G.V. Milovanović and the present writer to investigate the computational problems associated with integrals that contain Einstein's function as a weight function (and also a related function named after Fermi); see [Gautschi and Milovanović 1985]. Trying to find additional applications of such integrals, we came across the summation procedure outlined above, almost as an afterthought, but at the time could only provide somewhat contrived examples for its use. It is gratifying to see that the procedure indeed has much greater utility than originally thought, as it is capable of dealing with the whole class of series indicated in (A2.1), and indeed also with the companion class of series having alternating sign factors. Slowly convergent power series can be dealt with similarly by applying the Laplace transform technique only to the coefficients of the series; for an instance of this, see [Gautschi 1991b].

It remains now, however, to identify the function f in (A2.10) for series of type (A2.6), and to discuss the numerical evaluation of the integral on the right of (A2.10). This will be the topic of the next two sections.

A3. Special Function Theory

We need to express the general term of the series (A2.6) as a Laplace transform,

$$k^{\nu-1} \cdot \frac{1}{(k+a)^m} = (\mathcal{L}f)(k). \tag{A3.1}$$

To do this, we apply the well-known convolution theorem (cf. [Widder 1941, Theorem 12.1a]),

$$\mathcal{L}g \cdot \mathcal{L}h = \mathcal{L}g * h, \tag{A3.2}$$

where

$$(g*h)(t) = \int_0^t g(\tau)h(t-\tau)d\tau. \tag{A3.3}$$

Since,

$$k^{\nu-1} = \left(\mathcal{L}\frac{t^{-\nu}}{\Gamma(1-\nu)}\right)(k),$$

$$(k+a)^{-m} = \left(\mathcal{L}\frac{t^{m-1}}{(m-1)!}e^{-at}\right)(k),$$

we then find that (A3.1) holds with f given by

$$f(t) = \frac{1}{(m-1)!\Gamma(1-\nu)} \int_0^t e^{-a(t-\tau)}(t-\tau)^{m-1}\tau^{-\nu}d\tau.$$

After a change of variables, $\tau = tu$, this assumes the form

$$f(t) = \frac{t^{m-\nu}e^{-at}}{(m-1)!\Gamma(1-\nu)} \int_0^1 e^{atu}(1-u)^{m-1}u^{-\nu}du$$

and reveals the connection of f with Kummer's function $M(\alpha, \beta, z)$ involving parameters $\alpha = 1 - \nu$, $\beta = m + 1 - \nu$ and variable $z = at$ (cf. [Abramowitz and Stegun 1964, Eq. 13.2.1]). Indeed,

$$f(t) = t^{1-\nu}g_{m-1}(t;a,\nu), \tag{A3.4}$$

where

$$g_n(t;a,\nu)$$

$$= g_n(t) = \frac{t^n e^{-at}}{\Gamma(n+2-\nu)} M(1-\nu, n+2-\nu, at), \ n = 0, 1, 2, \ldots,$$

$$\text{Re}\, a \geq 0, \ \text{Im}\, a \geq 0, \ 0 < \nu < 1. \tag{A3.5}$$

For definiteness we assume that $a \neq 0$. (In the case $a = 0$, the series (A2.6) is expressible, and therefore easily computable, in terms of the Riemann zeta function.) The recurrence relation relative to the second parameter in Kummer's function [Abramowitz and Stegun 1964, Eq. 13.4.2] now immediately yields a three-term recurrence relation for the function g_n in (A3.5), namely

$$g_{n+1}(t) = \frac{1}{n+1}\left\{\left(t + \frac{n+1-\nu}{a}\right)g_n(t) - \frac{t}{a}g_{n-1}(t)\right\},$$

$$n = 0, 1, 2, \ldots,$$

$$g_{-1}(t) = \frac{t^{-1}}{\Gamma(1-\nu)}. \tag{A3.6}$$

It suffices, therefore, to focus attention on $g_0(t) = e^{-at}M(1-\nu, 2-\nu, at)/\Gamma(2-\nu)$, which is expressible [Abramowitz and Stegun 1964, Eq. 13.6.10] in terms of Tricomi's form of the incomplete gamma function [Abramowitz and Stegun 1964, Eq. 6.5.4],

$$g_0(t;a,\nu) = e^{-at}\gamma^*(1-\nu, -at), \tag{A3.7}$$

where

$$\gamma^*(\lambda, z) = \frac{z^{-\lambda}}{\Gamma(\lambda)} \int_0^z e^{-t} t^{\lambda-1} dt. \qquad (A3.8)$$

This is known to be an entire function of both its variables [Tricomi 1954, Ch. IV]. Consequently, g_0 is an entire function in all its variables, and so is g_n for each $n > 0$, considered as a function of t. Putting (A3.1), (A3.4) and (A2.10) together, we obtain

$$\sum_{k=1}^{\infty} \frac{k^{\nu-1}}{(k+a)^m} = \int_0^{\infty} t^{-\nu} \epsilon(t) \cdot g_{m-1}(t; a, \nu) dt,$$

$$\qquad (A3.9)$$

$$\mathrm{Re}\, a \geq 0, \ 0 < \nu < 1, \ m \geq 1.$$

(The formula (A3.9) holds also for $a = 0$, if one defines $g_0(t) = 1/\Gamma(2-\nu)$, $g_{n+1}(t) = t g_n(t)/(n+2-\nu)$, $n = 0, 1, 2, \ldots$.) We have managed to express the desired series as an integral containing the weight function $t^{-\nu}\epsilon(t)$, with ϵ given in (A2.9), and the factor g_{m-1}, an entire function of t. This almost begs for the use of weighted Gaussian quadrature. We will comply in the next section, where the issues involved will be further discussed. Here we conclude with two remarks.

We mentioned earlier that a shift in the summation index is easy to incorporate. Indeed, a shift in the variable of the Laplace transform corresponds to an exponential factor in the original function,

$$(\mathcal{L}f)(s+b) = (\mathcal{L}e^{-bt}f(t))(s). \qquad (A3.10)$$

Therefore,

$$(k+b)^{\nu-1} = \left(\mathcal{L} \frac{t^{-\nu}e^{-bt}}{\Gamma(1-\nu)} \right)(k),$$

and in place of (A3.4), one obtains

$$f(t) = e^{-bt} t^{1-\nu} g_{m-1}(t; a-b, \nu),$$

hence, by (A2.10),

$$\sum_{k=1}^{\infty} \frac{(k+b)^{\nu-1}}{(k+a)^m} = \int_0^{\infty} t^{-\nu}\epsilon(t) \cdot e^{-bt} g_{m-1}(t; a-b, \nu)dt,$$

(A3.11)

$$\operatorname{Re} a \geq 0, \ \operatorname{Re} b \geq 0, \ 0 < \nu < 1, \ m \geq 1.$$

A shift in the denominator index k, of course, can be absorbed by the constant a.

We finally remark that in the special case $\nu = \frac{1}{2}$, which is of prime interest to us, Tricomi's incomplete gamma function becomes Dawson's integral [Abramowitz and Stegun 1964, Eqs. 6.5.18, 7.1.17],

$$\gamma^* \left(\frac{1}{2}, -x^2\right) = \frac{2}{\sqrt{\pi} x} \int_0^x e^{t^2} dt,$$

and the function g_0 in (A3.7) becomes

$$g_0 \left(t; a, \frac{1}{2}\right) = \frac{2}{\sqrt{\pi}} \frac{F(\sqrt{at})}{\sqrt{at}},$$

(A3.12)

where

$$F(z) = e^{-z^2} \int_0^z e^{t^2} dt.$$

(A3.13)

This is a well-studied special function, related to the Gaussian error function, for which (almost) best uniform rational approximations on $[0,\infty]$ are known, yielding accuracies of up to twenty significant decimal digits [Cody et al. 1970]. It is easy, therefore, to compute this function to high accuracy, for any real argument.

Specializing (A3.11) to $\nu = \frac{1}{2}$, and letting $a = \alpha + 1$, $b = \alpha$, $m = 1$, yields

$$U(\alpha) = \frac{2}{\sqrt{\pi}} \int_0^{\infty} t^{-1/2}\epsilon(t)e^{-\alpha t}\frac{F(\sqrt{t})}{\sqrt{t}}dt,$$

(A3.14)

which identifies the mysterious function $U(\alpha)$ in (A1.4) as a Laplace transform, namely

$$U(\alpha) = (\mathcal{L}u)(\alpha), \qquad u(t) = \frac{2}{\sqrt{\pi}}t^{-1/2}\epsilon(t)\frac{F(\sqrt{t})}{\sqrt{t}}.$$

(A3.15)

The function u has a branch point at the origin and poles at integer multiples of $2\pi i$, but otherwise is regular analytic. The integral required in (A1.3) is obtained from (A3.14) by integrating with respect to α under the integral sign,

$$\int_0^\alpha U(\alpha)d\alpha = \frac{2\alpha}{\sqrt{\pi}} \int_0^\infty t^{-1/2}\epsilon(t) \cdot \frac{1 - e^{-\alpha t}}{\alpha t} \frac{F(\sqrt{t})}{\sqrt{t}}dt. \quad \text{(A3.16)}$$

This, too, is accessible to Gaussian quadrature.

A4. Numerical Implementation

It is time now to take a look at the numerical aspects of our basic formula (A3.9). We will concentrate on the case of particular interest to us, that is, $\nu = \frac{1}{2}$.

The first issue, then, is the numerical evaluation of g_{m-1} $(t; a, \frac{1}{2})$, the function that appears in the integrand of (A3.9). We naturally use the recurrence relation (A3.6) if $m > 1$, and (A3.12) otherwise. For real values of a, this poses no particular problems, since, as already mentioned, there are high-precision rational approximations available for Dawson's integral (A3.13). When the parameter a is complex, we are dealing essentially with the complex error function, for which efficient computer routines are also available (e.g., [Gautschi 1969; Poppe and Wijers 1990]), though not quite to the same high precision (approximately 10–14 decimals only).

Thus, it is not so much the integrand, as the integral itself,

$$I_{m-1}(a, \nu) = \int_0^\infty t^{-\nu}\epsilon(t) \cdot g_{m-1}(t; a, \nu)dt, \quad \text{(A4.1)}$$

that requires special scrutiny. One of its unpleasant features is the square root singularity at the origin (when $\nu = \frac{1}{2}$), another the fact that $\epsilon(t) = t/(e^t - 1)$ has a string of poles along the imaginary axis (at the integer multiples of $2\pi i$), and finally we must integrate to infinity. As we will see shortly, an additional difficulty looms underneath, and surfaces when a becomes large.

All difficulties, except the last, are swiftly overcome by treating

$$w(t) = t^{-1/2}\epsilon(t), \quad 0 < t < \infty, \tag{A4.2}$$

as a weight function and approximating the integral by Gaussian quadrature relative to this weight function. Thus,

$$I_{m-1}\left(a, \frac{1}{2}\right) \approx \sum_{r=1}^{n} \omega_r g_{m-1}\left(\tau_r; a, \frac{1}{2}\right), \tag{A4.3}$$

where $\omega_r = \omega_r^{(n)}(w)$, $\tau_r = \tau_r^{(n)}(w)$ are the weights and nodes of the n-point Gaussian quadrature formula for the weight function w. The computation of these formulae will be the subject of Section A5. Here we note that the quadrature process (A4.3) can be shown to converge as $n \to \infty$, since $g_{m-1}(t; a, \frac{1}{2})$ has at most polynomial growth as $t \to \infty$ and the moment problem for the weight function (A4.2) is determined (cf. [Freud 1971, Ch. III, Theorem 1.4]).

In Table A1 we show the n-point Gaussian approximations to $I_{m-1}(a, \frac{1}{2})$ for $m = 1$ and for selected values of a. (Note the Theodorus constant at the bottom of the third column!) Those for $m > 1$ are similar, though a bit more slowly convergent. As is evident from Table A1, convergence as $n \to \infty$ is quite fast when a is small or moderately large, but is slowing down conspicuously as a gets larger. (The reason for this slowdown is a peculiar behavior of the function $g_0(t; a, \frac{1}{2})$: as $a \to \infty$, it approaches the discontinuous function equal to $\frac{2}{\sqrt{\pi}}$ at $t = 0$ and zero for $t > 0$.) Before we show how to resolve this difficulty, it may be worthwhile indicating a somewhat simpler, if not necessarily more efficient, integration procedure.

We note that for large t,

$$\epsilon(t) = \frac{te^{-t}}{1 - e^{-t}} \sim te^{-t}, \quad t \to \infty,$$

which suggests writing (A4.1) in the form

$$I_{m-1}(a, \nu) = \int_0^\infty t^{-\nu} e^{-t} \cdot \frac{t}{1 - e^{-t}} g_{m-1}(t; a, \nu) dt. \tag{A4.4}$$

n	$a = .5$	$a = 1.$
5	2.1344163	1.8599
10	2.1344166429861	1.860025078
15	2.134416642986372611	1.86002507922117
20	2.134416642986372611	1.860025079221190306
25		1.8600250792211903071
30		1.8600250792211903072
35		
40		

n	$a = 2.$	$a = 4.$	$a = 8.$
5	1.537	1.19	.8
10	1.53967	1.217	.91
15	1.539680509	1.21826	.930
20	1.539680512350	1.218273	.9312
25	1.53968051235329	1.218274011	.93135
30	1.539680512353302010	1.21827401461	.931371
35	1.5396805123533020128	1.218274014668	.9313727
40	1.5396805123533020128	1.2182740146698	.93137291

Table A1: The n-point Gaussian approximations (A4.3) for $m = 1$, $a = .5, 1., 2., 4., 8.$

Hence, the new weight function

$$w^L(t) = t^{-\nu}e^{-t}, \quad 0 < t < \infty, \qquad (A4.5)$$

emerges, giving rise to classical (generalized) Gauss–Laguerre quadrature. The presence of the poles in the factor $t/(1 - e^{-t})$ now retards convergence somewhat, but interestingly enough, only when a is relatively small. By the time a reaches 2, Gauss–Laguerre quadrature in (A4.4) catches up with the more sophisticated quadrature in (A4.3), and indeed surpasses it as a is increased beyond 2, although both continue to struggle.

It thus remains to resolve the difficulty pertaining to large parameters a. (This is of no relevance to spirals, where $a = 1$, but should nevertheless be of intrinsic interest. We again assume $\nu = \frac{1}{2}$.) In this case we use a device, called *stratified summation* in [Gautschi 1991a], that consists of "layering" the summation in (A3.9) as follows:

$$k = \lambda + \kappa a_0, \quad a_0 = \lfloor a \rfloor, \quad a = a_0 + a_1, \qquad \text{(A4.6)}$$

where $\lfloor a \rfloor$ denotes the "floor of a," that is, the largest integer less than or equal to a, and $0 \leq a_1 < 1$. The summation is now carried out by letting κ run from 0 to ∞ for each $\lambda = 1, 2, \ldots, a_0$. Thus,

$$\sum_{k=1}^{\infty} \frac{k^{-1/2}}{(k+a)^m} = \sum_{\lambda=1}^{a_0} \sum_{\kappa=0}^{\infty} \frac{(\lambda + \kappa a_0)^{-1/2}}{(\lambda + \kappa a_0 + a_0 + a_1)^m}$$

$$= a_0^{-(m+1/2)} \sum_{\lambda=1}^{a_0} \left\{ \sum_{\kappa=1}^{\infty} \frac{(\kappa + \lambda/a_0)^{-1/2}}{(\kappa + 1 + (\lambda + a_1)/a_0)^m} \right.$$

$$\left. + \frac{(\lambda/a_0)^{-1/2}}{(1 + (\lambda + a_1)/a_0)^m} \right\}. \qquad \text{(A4.7)}$$

To the inner sum, we can now apply (A3.11), giving

$$\sum_{\kappa=1}^{\infty} \frac{(\kappa + \lambda/a_0)^{-1/2}}{(\kappa + 1 + (\lambda + a_1)/a_0)^m}$$

$$= \int_0^{\infty} t^{-1/2} \epsilon(t) e^{-(\lambda/a_0)t} g_{m-1}\left(t; 1 + a_1/a_0, \frac{1}{2}\right) dt. \qquad \text{(A4.8)}$$

Since the "effective" parameter in g_{m-1} is now between 1 and 2, and the coefficient λ/a_0 in the exponential is bounded by 1, either of the two quadrature schemes above ought to do quite well in computing the integral in (A4.8), hence the sum in (A4.7). It so happens that Gauss–Laguerre quadrature is now the faster of the two. The results obtained by this rule are shown in Table A2.

Although Gauss–Laguerre quadrature is not quite as efficient as the nonclassical quadrature in (A4.3) when a is small, convergence is then fast anyway, so that on the whole, it would be preferable to use Gauss–Laguerre if one had to choose

n	$a = 8.$	$a = 16.$
5	.931367	.694928
10	.9313729339	.69493171459
15	.931372934003102	.6949317146410448
20	.93137293400310387164	.69493171464104559014
25	.93137293400310387168	.69493171464104559016
30	.93137293400310387169	.69493171464104559016

n	$a = 32.$
5	.509924
10	.50992651699
15	.5099265170272109
20	.50992651702721134802
25	.50992651702721134803
30	.50992651702721134804

Table A2:   Approximations to the series in (A4.7) using n-point Gauss–Laguerre quadrature in (A4.8).

between one of the two. This is particularly so if many ν-values are involved. When one has to deal with series containing alternating sign factors, the choice is less clear, since then the appropriate weight function is $w(t) = 1/(e^t + 1)$ (known as "Fermi function" in solid-state physics; cf. [Gautschi and Milovanović 1985]), which has poles at odd multiples of πi, hence twice as close to the real axis as the poles of Einstein's function.

A5. Gaussian Quadrature Formulae and Their Computation

We call w a weight function on the interval (a, b) if w is non-negative and integrable on (a, b) and has finite moments of all orders,

$$\mu_s = \mu_s(w) = \int_a^b t^s w(t)dt, \ s = 0, 1, 2, \ldots, \quad (A5.1)$$

with $\mu_0 > 0$. The quadrature formula

$$\int_a^b f(t)w(t)dt = \sum_{r=1}^n \omega_r f(\tau_r) + R_n(f) \qquad (A5.2)$$

associated with the weight function w is called Gaussian if it is exact whenever f is a polynomial of degree $\leq 2n - 1$, that is, if

$$R_n(f) = 0 \text{ for all } f \in \mathbf{P}_{2n-1}. \qquad (A5.3)$$

This is best possible in the sense that no quadrature formula of the form (A5.2) exists having degree of exactness $> 2n - 1$. It is also well known that the Gaussian nodes $\tau_r = \tau_r^{(n)}(w)$ are mutually distinct and contained in (a, b), and that all weights $\omega_r = \omega_r^{(n)}(w)$ are positive (cf. [Davis and Rabinowitz 1984, §2.7]).

There is a close connection with orthogonal polynomials relative to the weight function w. This is a sequence of polynomials $\{\pi_k\}_{k=0}^\infty$, each $\pi_k(\cdot) = \pi_k(\cdot\,; w)$ monic of degree k and such that

$$(\pi_k, \pi_\ell) = 0 \text{ for all } \ell \neq k. \qquad (A5.4)$$

Here, $(\cdot, \cdot)$ is the inner product defined by

$$(u, v) = \int_a^b u(t)v(t)w(t)dt. \qquad (A5.5)$$

The nodes τ_r, indeed, are precisely the zeros of the nth-degree orthogonal polynomial $\pi_n(\cdot\,; w)$, while the weights ω_r can also be expressed (in various ways) in terms of these polynomials.

For computational purposes, however, it is more convenient to characterize these Gaussian quantities in terms of eigenvalues and eigenvectors of a symmetric tridiagonal matrix (cf. [Golub and Welsch 1969]). To arrive at such a characterization, we must recall that the orthogonal polynomials $\{\pi_k\}$ satisfy a three-term recurrence relation

$$\pi_{k+1}(t) = (t - \alpha_k)\pi_k(t) - \beta_k \pi_{k-1}(t), \quad k = 0, 1, 2, \ldots,$$
$$\qquad (A5.6)$$
$$\pi_{-1}(t) = 0, \quad \pi_0(t) = 1,$$

with coefficients $\alpha_k = \alpha_k(w) \in \mathbf{R}$, $\beta_k = \beta_k(w) > 0$ uniquely determined by the weight function w. The coefficient β_0 is arbitrary (since it multiplies $\pi_{-1} = 0$ in (A5.6)), but it is customary to define it by

$$\beta_0 = \beta_0(w) = \int_a^b w(t)dt \quad (= \mu_0). \qquad (A5.7)$$

For the normalized polynomials $\tilde{\pi}_k$ (satisfying $(\tilde{\pi}_k, \tilde{\pi}_k) = 1$), it then follows easily from (A5.6) that

$$t\tilde{\pi}_k(t) = \alpha_k \tilde{\pi}_k(t) + \sqrt{\beta_k}\tilde{\pi}_{k-1}(t) + \sqrt{\beta_{k+1}}\tilde{\pi}_{k+1}(t),$$

$$k = 0, 1, 2, \dots . \qquad (A5.8)$$

We now define the Jacobi matrix associated with the weight function w to be

$$J = J(w) = \mathrm{tri}(\alpha_0, \alpha_1, \dots; \sqrt{\beta_1}, \sqrt{\beta_2}, \dots), \qquad (A5.9)$$

the infinite symmetric tridiagonal matrix having the coefficients $\alpha_0(w), \alpha_1(w), \dots$ down the main diagonal, and the square roots $\sqrt{\beta_1(w)}, \sqrt{\beta_2(w)}, \dots$ down each side diagonal. The truncated Jacobi matrix,

$$J_n = J_n(w) = [J(w)]_{n \times n}, \qquad (A5.10)$$

is the top left $n \times n$ section of $J(w)$.

The first n relations in (A5.8) can now be expressed in vector form as

$$t\tilde{\pi}(t) = J_n \tilde{\pi}(t) + \sqrt{\beta_n}\tilde{\pi}_n(t)e_n, \qquad (A5.11)$$

where $\tilde{\pi}(t) = [\tilde{\pi}_0(t), \tilde{\pi}_1(t), \dots, \tilde{\pi}_{n-1}(t)]^T$ and $e_n = [0, 0, \dots, 0, 1]^T$ are vectors in $\mathbf{R}^n$. Since τ_r is a zero of $\tilde{\pi}_n$, it follows from (A5.11) immediately that

$$\tau_r \tilde{\pi}(\tau_r) = J_n \tilde{\pi}(\tau_r), \quad r = 1, 2, \dots, n, \qquad (A5.12)$$

that is, *the Gaussian nodes* $\tau_r = \tau_r^{(n)}(w)$ *are the eigenvalues of* $J_n = J_n(w)$, *and* $\tilde{\pi}(\tau_r)$ *the corresponding eigenvectors.* Note indeed that $\parallel \tilde{\pi}(\tau_r) \parallel \neq 0$, since the first component $\tilde{\pi}_0$ is a positive number, namely

$$\tilde{\pi}_0 = \mu_0^{-1/2}. \tag{A5.13}$$

It remains to express the Gaussian weights in terms of the eigenvectors. To do so, let v_r be the normalized eigenvectors,

$$J_n v_r = \tau_r v_r, \quad v_r^T v_r = 1, \tag{A5.14}$$

so that

$$v_r = \frac{1}{\sqrt{\tilde{\pi}(\tau_r)^T \tilde{\pi}(\tau_r)}} \tilde{\pi}(\tau_r) = \left(\sum_{\mu=1}^{n} \tilde{\pi}_{\mu-1}^2(\tau_r) \right)^{-1/2} \tilde{\pi}(\tau_r).$$

Comparing the first component on each side, and squaring, one obtains by virtue of (A5.13), (A5.7),

$$\frac{1}{\sum_{\mu=1}^{n} \tilde{\pi}_{\mu-1}^2(\tau_r)} = \beta_0 v_{r,1}^2, \quad r = 1, 2, \ldots, n. \tag{A5.15}$$

Here, $v_{r,1}$ denotes the first component of v_r. On the other hand, letting $f(t) = \tilde{\pi}_{\mu-1}(t)$ in (A5.2), one gets by orthogonality, again using (A5.13),

$$\beta_0^{1/2} \delta_{\mu-1,0} = \sum_{r=1}^{n} \tilde{\pi}_{\mu-1}(\tau_r) \omega_r \quad (\delta_{\mu-1,0} = \text{Kronecker delta}),$$

or, in matrix form,

$$P\omega = \beta_0^{1/2} e_1, \tag{A5.16}$$

where $P \in \mathbf{R}^{n \times n}$ is the matrix of eigenvectors, $\omega \in \mathbf{R}^n$ is the vector of Gauss weights, and $e_1 = [1, 0, \ldots, 0]^T \in \mathbf{R}^n$. Since the columns of P are orthogonal, we have

$$P^T P = D, \quad D = \text{diag}\,(d_1, d_2, \ldots, d_n), \quad d_r = \sum_{\mu=1}^{n} \tilde{\pi}_{\mu-1}^2(\tau_r).$$

Now multiplying (A5.16) from the left by P^T gives

$$D\omega = \beta_0^{1/2} P^T e_1 = \beta_0^{1/2} \cdot \beta_0^{-1/2} e = e, \quad e = [1, 1, \ldots, 1]^T.$$

Therefore, $\omega = D^{-1}e$, that is,

$$\omega_r = \frac{1}{\sum_{\mu=1}^{n} \tilde{\pi}_{\mu-1}^2(\tau_r)}, \quad r = 1, 2, \ldots, n.$$

Comparing this with (A5.15), we get the desired result,

$$\omega_r = \beta_0 v_{r,1}^2, \quad r = 1, 2, \ldots, n. \qquad (A5.17)$$

Thus, *the Gaussian weights* $\omega_r = \omega_r^{(n)}(w)$ *are the squares of the first component of the normalized eigenvectors of* $J_n = J_n(w)$ *multiplied by* β_0 (cf. (A5.7)).

The computation of Gaussian quadrature formulae is thus reduced to solving an eigenvalue/eigenvector problem for the symmetric tridiagonal matrix $J_n = J_n(w)$. This is a standard problem in numerical linear algebra that can be solved very efficiently by the QR (or QL) algorithm with carefully selected shifts (see, e.g., [Parlett 1980, §§8.9–8.11]).

We must not forget, however, that this process assumes that the Jacobi matrix $J = J(w)$ is explicitly known, that is, that we know the recursion coefficients in (A5.6). Fortunately, this is the case for all classical orthogonal polynomials. Those of interest to us – the generalized Laguerre polynomials – corresponding to the weight function $w(t) = t^{-\nu}e^{-t}$ on $(0, \infty)$, indeed have particularly simple coefficients, namely

$$\alpha_k(w) = 2k + 1 - \nu, \quad k \geq 0;$$

$$\beta_0(w) = \Gamma(1 - \nu), \quad \beta_k(w) = k(k - \nu), \quad k \geq 1, \qquad (A5.18)$$

$$(w(t) = t^{-\nu}e^{-t}).$$

Thus, it is a simple matter to generate the associated Gauss–Laguerre quadrature formulae for any $\nu \in (0, 1)$ and any $n = 1, 2, 3. \ldots$

Computing the Jacobi matrix for nonclassical weight functions, such as $w(t) = t^{-\nu}\epsilon(t)$ (cf. (A3.9)), is a much harder problem. Nevertheless, a number of techniques are available (for a discussion of these, we must refer to the literature, e.g., [Gautschi 1990]). The most appropriate for our purposes here consists of approximating the respective inner product (A5.5) by a discrete inner product,

$$(u, v) \approx (u, v)_N, \quad (u, v)_N = \sum_{k=1}^{N} w_k^{(N)} u(t_k^{(N)}) v(t_k^{(N)}), \quad (A5.19)$$

and then approximating the desired recursion coefficients by

$$\alpha_k \approx \alpha_k^{(N)}, \quad \beta_k \approx \beta_k^{(N)}, \quad (A5.20)$$

the recursion coefficients belonging to the polynomials orthogonal with respect to the discrete inner product $(u, v)_N$. (These can be computed in various ways.) The process, if properly implemented, can be made to converge in the sense that

$$\alpha_k = \lim_{N \to \infty} \alpha_k^{(N)}, \quad \beta_k = \lim_{N \to \infty} \beta_k^{(N)} \quad (A5.21)$$

for any fixed k.

As an example, for the weight function $w(t) = t^{-\nu} \epsilon(t)$, we can use, similarly as in (A4.4), Gauss–Laguerre quadrature,

$$\begin{aligned}
(u, v) &= \int_0^\infty t^{-\nu} \frac{t}{e^t - 1} u(t) v(t) dt \\
&= \int_0^\infty t^{-\nu} e^{-t} \frac{t}{1 - e^{-t}} u(t) v(t) dt \\
&\approx \sum_{k=1}^{N} \omega_k \frac{\tau_k}{1 - e^{-\tau_k}} u(\tau_k) v(\tau_k),
\end{aligned}$$

to obtain the discretization (A5.19) with

$$w_k^{(N)} = \omega_k \frac{\tau_k}{1 - e^{-\tau_k}}, \quad t_k^{(N)} = \tau_k.$$

The quadrature nodes $\tau_k = \tau_k^{(N)}$ and weights $\omega_k = \omega_k^{(N)}$ of the Gauss-Laguerre formula are computed as discussed above.

References

[Abramowitz and Stegun 1964] M. Abramowitz and I. A. Stegun (eds.), *Handbook of Mathematical Functions*, NBS Applied Math.

Ser., vol. 55. Washington, D.C.: U.S. Government Printing Office, 1964.

[Cody et al. 1970] W. J. Cody, K. A. Paciorek and H. C. Thacher, Jr., "Chebyshev approximations for Dawson's integral," *Math. Comp.* **24** (1970), 171–178.

[Davis and Rabinowitz 1984] P. J. Davis and P. Rabinowitz, *Methods of Numerical Integration*, 2nd ed. Orlando: Academic Press, 1984.

[Freud 1971] G. Freud, *Orthogonal Polynomials*. New York: Pergamon Press, 1971.

[Gautschi 1969] W. Gautschi, "Algorithm 363 – Complex error function," *Comm. ACM* **12** (1969), 635.

[Gautschi 1990] W. Gautschi, "Computational aspects of orthogonal polynomials," in: *Orthogonal Polynomials – Theory and Practice* (P. Nevai, ed.), NATO ASI Series, Series C: Mathematical and Physical Sciences, vol. 294, pp. 181–216. Dordrecht: Kluwer, 1990.

[Gautschi 1991a] W. Gautschi, "A class of slowly convergent series and their summation by Gaussian quadrature," *Math. Comp.* **57** (1991), 309–324.

[Gautschi 1991b] W. Gautschi, "On certain slowly convergent series occurring in plate contact problems," *Math. Comp.* **57** (1991), 325–338.

[Gautschi and Milovanović 1985] W. Gautschi and G. V. Milovanović, "Gaussian quadrature involving Einstein and Fermi functions with an application to summation of series," *Math. Comp.* **44** (1985), 177–190.

[Golub and Welsch 1969] G. H. Golub and J. H. Welsch, "Calculation of Gauss quadrature rules," *Math. Comp.* **23** (1969), 221–230. Loose microfiche suppl. A1–A10.

[Henrici 1984] P. Henrici, *Applied and Computational Complex Analysis*, vol. 1. New York: Wiley, 1984.

[Parlett 1980] B. N. Parlett, *The Symmetric Eigenvalue Problem*. Englewood Cliffs: Prentice-Hall, 1980.

[Poppe and Wijers 1990] G. P. M. Poppe and C. M. J. Wijers, "Algorithm 680 – Evaluation of the complex error function," *ACM Trans. Math. Software* **16** (1990), 47.

[Tricomi 1954] F. G. Tricomi, *Funzioni Ipergeometriche Confluenti*. Roma: Edizioni Cremonese, 1954.

[Widder 1941] D. V. Widder, *The Laplace Transform*. Princeton: University Press, 1941.

Supplement B

The Dynamics of the Theodorus Spiral

A. Iserles

B1. The One-Dimensional Theodorus Spiral

Let A and B be two arbitrary complex $d \times d$ matrices. The *generalized Theodorus spiral* is defined by

$$\mathbf{z}_{n+1} = A z_n + B \frac{z_n}{\|z_n\|}, \qquad \text{(B1.1)}$$

with an initial vector $\mathbf{z}_0 \in \mathcal{C}^d$. Here $\| \cdot \|$ is, in principle, an arbitrary vector norm. However, throughout these notes, we confine our attention to the Euclidean norm

$$\|\mathbf{x}\| = \left\{ \sum_{j=1}^{d} |x_j|^2 \right\}^{1/2}.$$

Even this is, actually, too general: the dynamics of (B1.1) is nontrivial even in the simplest, one-dimensional case – nontrivial enough to justify a whole introductory section!

Thus, let us consider

$$z_{n+1} = az_n + b\frac{z_n}{|z_n|}, \tag{B1.2}$$

where $a, b, z_0 \in \mathcal{C}$, $b \neq 0$. Let $\tilde{z}_n := z_n/|b|$. Therefore,

$$\tilde{z}_{n+1} = a\tilde{z}_n + e^{i \arg b}\frac{\tilde{z}_n}{|\tilde{z}_n|}. \tag{B1.3}$$

This is almost identical to (B1.2), except that the second constant on the right is of unit modulus. Both equations are equivalent: if we know what "happens" in (B1.3), we can always translate the answer back to the language of $\{z_n\}_{n=0}^{\infty}$ by rescaling. Consequently, we may assume with no loss of generality that $|b| = 1$ in (B1.2).

Let $a = \rho e^{i\psi}$, $b = e^{i\varphi}$, where $\rho > 0$, and suppose that each z_n has the polar representation $z_n = r_n e^{i\theta_n}$, $n = 0, 1, \ldots$. Substitution in (B1.2) affirms that

$$r_{n+1}e^{i\theta_{n+1}} = \rho r_n e^{i(\psi+\theta_n)} + e^{i(\varphi+\theta_n)}, \quad n = 0, 1, \ldots.$$

We take absolute values on both sides:

$$r_{n+1} = \left\{\rho^2 r_n^2 + 2\rho r_n \cos\gamma + 1\right\}^{1/2}, \quad n = 0, 1, \ldots, \tag{B1.4}$$

where $\gamma := \psi - \varphi$.

What are the possible asymptotic values of (B1.4)? Given an arbitrary d-dimensional functional iteration scheme

$$\mathbf{z}_0 \in \mathcal{C} \text{ given,}$$
$$\mathbf{z}_{n+1} = \mathbf{f}(\mathbf{z}_n), \quad n = 0, 1, \ldots,$$

the possible limits are the fixed points, that is, zeros of the equation $\mathbf{z} = \mathbf{f}(\mathbf{z})$. Moreover, suppose that $\mathbf{f}$ is differentiable at a fixed point $\hat{\mathbf{z}}$ and that $J(\mathbf{z}) = \partial\mathbf{f}(\hat{\mathbf{z}})/\partial\mathbf{z}$. Then $\hat{\mathbf{z}}$ is attractive – that is, there exists an open neighborhood $\mathcal{U}$ of $\hat{\mathbf{z}}$ such that

$\mathbf{z}_0 \in \mathcal{U}$ implies $\lim_{n\to\infty} \mathbf{z}_n = \hat{\mathbf{z}}$ – if and only if all the eigenvalues of $J(\hat{\mathbf{z}})$ reside in the open unit disk $|z| < 1$. In our case,

$$f(r) = \{\rho^2 r^2 + 2\rho r \cos\gamma + 1\}^{1/2},$$

and there are two fixed points, namely

$$r_\pm = \frac{1}{1-\rho^2}\left(\rho\cos\gamma \pm \left\{1 - \rho^2\sin^2\gamma\right\}^{1/2}\right),$$

provided that $\rho \neq 1$ and $\rho|\sin\gamma| \leq 1$. If $\rho = 1$, we have just one fixed point,

$$r_+ = -\frac{1}{2\cos\gamma}$$

(as long as $\cos\gamma \neq 0$). However, of course any limit point of (B1.2) must be positive: Recall that r_n is the absolute value of z_n! This reduces the options, since

$$0 < \rho < 1 \qquad \Longrightarrow \qquad r_- < 0 < r_+;$$
$$\rho = 1 \qquad \Longrightarrow \qquad r_+ > 0 \text{ if and only if } \cos\gamma < 0,$$
$$1 < \rho < \frac{1}{|\sin\gamma|} \qquad \Longrightarrow \qquad \text{both } r_- > 0 \text{ and } r_+ > 0$$
$$\text{if and only if } \cos\gamma < 0.$$

Moreover,

$$\frac{df(r_\pm)}{dr} = \rho^2\sin^2\gamma \pm \rho\cos\gamma\left\{1 - \rho^2\sin^2\gamma\right\}^{1/2}.$$

Consequently, attractivity takes place if and only if $\rho \leq 1$ and only the fixed point r_+ can be attractive.

Each attractive fixed point r_+ comes with its own *basin of attraction*, namely an open set $\mathcal{B}(r_+) \in (0, \infty)$ such that

$$r_0 \in \mathcal{B}(r_+) \Rightarrow \lim_{n\to\infty} r_n = r_+.$$

Provided that $\gamma \neq 0 \bmod \pi$, we have

$$\frac{d^2 f(r)}{dr^2} = \frac{\rho^2 \sin^2\gamma}{f^3(r)} > 0,$$

hence f is strictly convex in $(0, \infty)$. This is enough to argue that $\mathcal{B}(r_+) = (0, \infty)$, by a standard geometrical argument. Suppose that $r_0 > r_+$. Then, by virtue of convexity, $r_+ < r_1 < r_0$ and, by induction,

$$r_+ < \cdots < r_n < r_{n-1} < \cdots < r_1 < r_0$$

for all n. We have a monotonically descending sequence, bounded from below by r_+. It must converge and the only possible limit is r_+! The proof for $0 < r_0 < r_+$ is a mirror image: We now obtain a monotonically ascending sequence. Hence all $(0, \infty)$ gives convergence. Actually, we have much more than this! The convergence is monotone: each consecutive iteration leads us closer to the fixed point.

The case $\gamma = 0 \bmod 2\pi$ is even easier, since now $f(r) = \rho r + 1$, a linear map. Note that $\gamma = 0$ cannot coexist with $\rho = 1$, since attractivity, in tandem with $\rho = 1$, requires $\cos \gamma < 0$. Hence $\rho < 1$. We can easily derive the explicit form of the nth iterate,

$$r_n = \frac{1}{1-\rho} + \rho^n \left(r_0 - \frac{1}{1-\rho} \right) \to \frac{1}{1-\rho}.$$

Again $\mathcal{B}(r_+) = (0, \infty)$, and again the sequence $\{r_n\}_{n=0}^{\infty}$ converges monotonically.

Finally, let $\gamma = \pi \bmod 2\pi$. We have now $f(r) = |\rho r - 1|$, a convex, nonnegative, piecewise-linear map that vanishes at $r = \rho^{-1}$. Let us assume first that $\rho < 1$. If $r_0 \in (0, \rho^{-1})$, then, explicitly,

$$r_n = 1 - \rho + \rho^2 - \cdots + (\rho)^{n-1} + (-\rho)^n r_0$$

$$= \frac{1}{1+\rho} + (-\rho)^n \left(r_0 - \frac{1}{1+\rho} \right) \to \frac{1}{1+\rho} = r_+.$$

Note, however, that the convergence is no longer monotone and the r_n's oscillate about r_+. Next, let r_0 belong to an interval of the form

$$\mathcal{I}_m := \left(\frac{\rho^{-m} - 1}{1 - \rho}, \frac{\rho^{-m-1} - 1}{1 - \rho} \right)$$

for some natural number m. It is straightforward to observe that, as long as $m \geq 1$, $r_1 \in \mathcal{I}_{m-1}$ and, in general, $r_n \in \mathcal{I}_{m-n}$ for $n = 0, \ldots, m$. Moreover, $r_{m+1} < \rho^{-1}$. Thus, after $m + 1$ monotone steps, we reach $(0, \rho^{-1})$, switching into a convergent oscillatory regime.

Next, we need to examine the points $\iota_m := (\rho^{-m} - 1)/(1 - \rho)$, $m = 0, 1, \ldots$. If $r_0 = \iota_m$, say, then $r_n = \iota_{m-n}$ for $n = 0, \ldots, m$, whereas $r_{m+1} = 0$ and $r_{m+2} = 1$. The "jump" through zero is, actually, accompanied by a little bit of mathematical licence: clearly, the quotient in (B1.2) is not defined when $z_n = 0$. An easy way out is to single out all the points $\{\iota_m\}_{m=0}^{\infty}$ as the "basin of attraction of ill-definition." It is more interesting by far to allow a jump through zero as a limiting case of $0 < |z_n| \ll 1$. This extends the "interesting" range of r_n's to $[0, \infty)$.

Having "bounced" back from zero, there are two possibilities: Either $1 \in (0, \infty) \setminus \{\iota_0, \iota_1, \ldots\}$ and we converge to r_+ or, alternatively, we land on some ι_k and are destined to repeat the journey through 0 indefinitely in a periodic orbit. Let us examine the second possibility. To "land" exactly on ι_k, it must be true that $\iota_k = 1$, thus $R_k(\rho) = 0$, where

$$R_k(\rho) = \rho^k - \sum_{j=0}^{k-1} \rho^j.$$

Clearly, $R_k(\rho) < 0$ for all $k = 1, 2, \ldots$, and no periodic orbits are possible for $\rho < 1$. However, recall that we restricted the range of ρ purely to ensure the attractivity of r_+. As far as periodic orbits are concerned, there is absolutely no need to confine ourselves to $\rho \leq 1$, and any positive zero of R_k will do. But $R_k(1) = 1 - k \leq 0$, $R_k(2) = 1$, and it follows that for every $k \geq 1$, there exists a value of $\rho \in [0, 2)$ that leads to a $(k + 1)$-periodic orbit. We leave it as an easy exercise to verify that for every $k \geq 0$, there exists a single positive zero ρ_k of R_k and that

$$1 = \rho_1 < \rho_2 < \cdots < \rho_n \overset{n \to \infty}{\longrightarrow} 2$$

(you may start by noticing that $R_{k+1}(\rho) = \rho R_k(\rho) - 1$). Thus, there always exists a "rogue" $\rho = |a| \geq 1$ that produces periodic sequences for appropriate r_0's.

Our analysis of the behavior of (B1.4) near its fixed points is complete, except for the case $\rho = 1$, $\cos \gamma = -1$. This can be treated by letting $\rho \uparrow 1$. Hence, if r_0 is noninteger, we converge to $r_+ = \frac{1}{2}$, whereas integer r_0 leads to the 2-cycle $\{0, 1\}$.

Of course, the dynamics of the modulus is only half of the story, and we need to look at the argument as well. We have

$$r_{n+1} \cos \theta_{n+1} = \rho r_n \cos(\psi + \theta_n) + \cos(\varphi + \theta_n),$$

$$r_{n+1} \sin \theta_{n+1} = \rho r_n \sin(\psi + \theta_n) + \sin(\varphi + \theta_n),$$

or, in a matrix notation,

$$\begin{bmatrix} \cos \theta_{n+1} \\ \sin \theta_{n+1} \end{bmatrix} = \begin{bmatrix} \alpha_n & -\beta_n \\ \beta_n & \alpha_n \end{bmatrix} \begin{bmatrix} \cos \theta_n \\ \sin \theta_n \end{bmatrix}, \quad n = 0, 1, \ldots,$$

where

$$\alpha_n = \frac{\rho r_n \cos \psi + \cos \varphi}{r_{n+1}}, \quad \beta_n = \frac{\rho r_n \sin \psi + \sin \varphi}{r_{n+1}}.$$

Note that $\alpha_n^2 + \beta_n^2 \equiv 1$, by virtue of (B1.4). Thus, for every $n = 0, 1, \ldots$, there exists a unique $\tau_n \in (-\pi, \pi]$ such that $e^{i \tau_n} = \alpha_n + i \beta_n$, and therefore $\theta_{n+1} = \theta_n + \tau_n \mod 2\pi$.

Let us consider first the case of $r_0 = r_+$. In other words, we commence from the "right" radius, but not necessarily from the "right" argument. Since now $r_n \equiv r_+$, it follows that $\tau_n \equiv \tau$ and $\theta_n = n\tau + \theta_0$. Thus, if $\tau \mod \pi$ is rational, $\tau = 2\pi K/L$, where $L \geq 1$ and the integers K, L are relatively prime, then $\theta_{n+L} = \theta_n \mod 2\pi$ for all $n = 0, 1, \ldots$. We obtain an L-periodic orbit, hopping along shifted Lth roots of unity. On the other hand, if τ is irrational modulo π, then the sequence $\{z_n\}_{n=0}^{\infty}$ is ergodic, equally distributed on the circle of radius r_+.

The more interesting case is $r_0 \in \mathcal{B}(r_+) \backslash \{r_+\}$. Since $r_n \to r_+$, obviously $\alpha_n \to \alpha$, $\beta_n \to \beta$, such that $\alpha + i\beta = e^{i\tau}$. Since

$\theta_{n+1} = \theta_n + \tau_n \bmod 2\pi$, we can express the angle θ_n in terms of the local revolutions τ_ℓ,

$$\theta_n = \sum_{\ell=0}^{n-1} \tau_\ell.$$

We write $\nu_+ := f'(r_+)$ and observe that, by virtue of attractivity, $|\nu_+| < 1$. Given any $\varepsilon > 0$, there exists N such that $|r_n - r_+| < \varepsilon$ for all $n \geq N$. Consequently,

$$r_n = r_+ + \nu_+^{n-N}\varepsilon + \mathcal{O}(\varepsilon^2), n \geq N,$$

and this, in turn, implies

$$\tau_n = \tau + \nu_+^{n-N}\delta + \mathcal{O}(\delta^2),$$

where $|\delta| \leq C\varepsilon$ for a constant $C > 0$, uniformly in ε and n. Thus,

$$\theta_n = \theta_N + \sum_{k=N}^{n-1} \tau_n = \theta_N + (n-N)\tau + \delta\frac{1 - \nu_+^{n-N}}{1 - \nu_+} + \mathcal{O}(\delta^2) \quad \bmod 2\pi.$$

Recall that $|\nu_+| < 1$. It follows that there exists θ^* such that

$$\theta_n = \theta^* + n\tau + o(1), \quad n \to \infty \quad \bmod 2\pi. \tag{B1.5}$$

Seemingly, the asymptotic behavior is the same as for $|z_0| = r_+$. This is an illusion! For previously everything depended on the rationality of τ/π, a property that is easily overwhelmed by the $o(1)$ term in (B1.5).

To recap, the modulus in (B1.2) converges whenever either $|a| < 1$ or $|a| = 1$, $\cos(\psi - \varphi) < 0$. However, the sequence $\{z_n\}$ itself converges only if $\tau = 0 \bmod 2\pi$. Otherwise it is either eventually periodic or (and this is a much more likely state of affairs) it is asymptotically ergodic on the circle $|z| = r_+$.

Observe that we considered the *magnitude* and the *orientation* of z_n separately. Both their analysis and the pattern of their behavior are different: it is far more difficult to analyze the magnitude, whereas the orientation is almost trivial. On the other hand, the behavior (as opposed to its analysis) of magnitude is straightforward and almost boring, whereas orientation is allowed all the games and fun. We emphasise this point, since the dichotomy persists in any number of dimensions, as is evidenced by the remainder of this article.

B2. The General Case

The one-dimensional Theodorus spiral being nontrivial, it should come as no surprise that d dimensions present a more formidable challenge. Suppose that $\hat{\mathbf{z}} = \lim_{n \to \infty} \mathbf{z}_n$ exists and (B1.1) has a fixed point. Thus,

$$\hat{\mathbf{z}} = \left(A + \frac{1}{\|\hat{\mathbf{z}}\|} B \right) \hat{\mathbf{z}}.$$

In other words, there exists a constant $c > 0$ such that 1 is an eigenvalue of $A + cB$, while $\hat{\mathbf{z}}$ is a member of $\mathrm{Ker}\,(A + cB - I) \setminus \{\mathbf{0}\}$, scaled so that $c\|\hat{\mathbf{z}}\| = 1$. Of course, not for every pair of matrices $\{A, B\}$ there exists such $c > 0$.

Suppose first that A and B commute and possess a full set of (joint) eigenvectors $\{\mathbf{v}_1, \ldots, \mathbf{v}_d\}$, with the corresponding eigenvalues

$$\sigma(A) = \{\lambda_1, \ldots, \lambda_d\}, \qquad \sigma(B) = \{\mu_1, \ldots, \mu_d\}.$$

It is necessary and sufficient for the existence of a fixed point that

$$\frac{\mu_p}{1 - \lambda_p} > 0$$

for some index $p \in \{1, 2, \ldots, d\}$. Let us assume that this is, indeed, the case, and that it occurs for just one index p. Then $\hat{\mathbf{z}} := \mathbf{v}_p$, normalized so that $\|\hat{\mathbf{z}}\| = \mu_p/(1 - \lambda_p)$.

Of course, we have already seen that the existence of a fixed point is only part of the story and that it is essential to check attractivity. For that purpose, we check, in tune with Section 1, whether the eigenvalues of the Jacobian matrix of the mapping at $\hat{\mathbf{z}}$ are all inside the unit disk. It is easy to verify that the Jacobian is given explicitly by

$$J = A + \frac{1}{\|\hat{\mathbf{z}}\|}B - \frac{1}{\|\hat{\mathbf{z}}\|^3}B\hat{\mathbf{z}}\hat{\mathbf{z}}^*. \tag{B2.1}$$

Set $\omega_\ell := c^3\mu_p\mathbf{v}_p^*\mathbf{v}_\ell$, $\ell = 1,\ldots,d$. Hence, for all ℓ in $\{1,\ldots,d\}$,

$$J\mathbf{v}_\ell = (\lambda_\ell + c\mu_\ell)\mathbf{v}_\ell - \omega_\ell\mathbf{v}_p. \tag{B2.2}$$

Since $\omega_p = c\mu_p$, it follows that $\mathbf{v}_p$ is an eigenvector of J, with the eigenvalue λ_p. Moreover, if $\mathbf{v}_\ell$ is orthogonal to $\mathbf{v}_p$ (as will be the case with all $\ell \neq p$ if A and B are symmetric), then $\omega_\ell = 0$ and $\lambda_\ell + c\mu_\ell \in \sigma(J)$. To obtain the remaining eigenvalues, we consider $\ell \neq p$, $\omega_\ell \neq 0$. Our contention is that

$$\mathbf{w} := \frac{\lambda_p - \lambda_\ell - c\mu_\ell}{\omega_\ell}\mathbf{v}_\ell + \mathbf{v}_p$$

is an eigenvector, with the corresponding eigenvalue $\lambda_\ell + c\mu_\ell$. This is a straightforward consequence of (B2.2), since

$$J\mathbf{w} = \frac{\lambda_p - \lambda_\ell - c\mu_\ell}{\omega_\ell}\left((\lambda_\ell + c\mu_\ell)\mathbf{v}_\ell - \omega_\ell\mathbf{v}_p\right) + \lambda_p\mathbf{v}_p$$

$$= (\lambda_\ell + c\mu_\ell)\frac{\lambda_p - \lambda_\ell - c\mu_\ell}{\omega_\ell}\mathbf{v}_\ell + (\lambda_\ell + c\mu_\ell)\mathbf{v}_p = (\lambda_\ell + c\mu_\ell)\mathbf{w}.$$

Except for the special case when $\lambda_\ell + c\mu_\ell = \lambda_p$, $\omega_\ell \neq 0$, for some ℓ, we recover all the eigenvalues of J,

$$\sigma(J) = \{\lambda_p\}\bigcup\left\{\lambda_\ell + \frac{\mu_\ell}{\mu_p}(1 - \lambda_p) : \ell = 1,\ldots,d,\ \ell \neq p\right\}$$

(recall that $c = (1 - \lambda_p)/\mu_p$). Thus, the conditions for attractivity are

$$\frac{1-\lambda_p}{\mu_p} > 0 \quad \text{(so that the fixed point exists)};$$
$$|\lambda_p| < 1;$$

and

$$\left|\lambda_\ell + \frac{\mu_\ell}{\mu_p}(1 - \lambda_p)\right| < 1, \quad \ell = 1,\ldots,d,\ \ell \neq p.$$

For our next example of a fixed point, we choose another instance of "extreme" behavior, letting both A and B be $d \times d$ Jordan blocks,

$$A = \begin{bmatrix} \lambda & 1 & 0 & \cdots & 0 \\ 0 & \lambda & 1 & \ddots & \vdots \\ \vdots & \ddots & \ddots & \ddots & 0 \\ \vdots & & \ddots & \lambda & 1 \\ 0 & \cdots & \cdots & 0 & \lambda \end{bmatrix}, \qquad B = \begin{bmatrix} \mu & 1 & 0 & \cdots & 0 \\ 0 & \mu & 1 & \ddots & \vdots \\ \vdots & \ddots & \ddots & \ddots & 0 \\ \vdots & & \ddots & \mu & 1 \\ 0 & \cdots & \cdots & 0 & \mu \end{bmatrix}.$$

Again, we denote a fixed point by $\hat{\mathbf{z}}$. Its components obey the equations

$$\hat{z}_\ell = \lambda \hat{z}_\ell + \hat{z}_{\ell+1} + \frac{1}{\|\hat{\mathbf{z}}\|}(\mu \hat{z}_\ell + \hat{z}_{\ell+1}), \quad \ell = 1, \ldots, d-1, \quad \text{(B2.3)}$$

$$\hat{z}_d = \left(\lambda + \frac{\mu}{\|\hat{\mathbf{z}}\|}\right)\hat{z}_d. \tag{B2.4}$$

It follows from (B2.4) that either $\hat{z}_d = 0$ or $\|\hat{\mathbf{z}}\| = \mu/(1-\lambda)$. In the first case we obtain in (B2.3)

$$\hat{z}_{d-1} = \left(\lambda + \frac{\mu}{\|\hat{\mathbf{z}}\|}\right)\hat{z}_{d-1}.$$

Thus, by the same token, either $\hat{z}_{d-1} = 0$ or $\|\hat{\mathbf{z}}\| = \mu/(1-\lambda)$. We continue by induction, proving that either $\hat{z}_\ell = 0$ for all $\ell = 2, \ldots, d$ or $\|\hat{\mathbf{z}}\| = \mu/(1-\lambda)$.

Let us suppose that the second alternative is true. Substitution into (B2.3) gives

$$\hat{z}_\ell = \hat{z}_\ell + \left(1 + \frac{1-\lambda}{\mu}\right)\hat{z}_{\ell+1},$$

hence

$$\left(1 + \frac{1-\lambda}{\mu}\right)\hat{z}_{\ell+1} = 0.$$

Since

$$\frac{1 - \lambda}{\mu} = \frac{1}{\|\hat{\mathbf{z}}\|} > 0,$$

it follows that $\hat{z}_{\ell+1} = 0$. But this is true for all $\ell = 1, \ldots, d-1$, and again we have $\hat{z}_\ell = 0$, $\ell = 2, \ldots, d$. Either way, $\hat{\mathbf{z}}$ is a scalar multiple of the first unit vector $\mathbf{e}_1$. Moreover, it follows from (B2.3) (for $\ell = 1$) that $\|\hat{\mathbf{z}}\| = \mu/(1 - \lambda)$, hence

$$\hat{\mathbf{z}} = \frac{\mu}{1 - \lambda} \mathbf{e}_1.$$

For this, of course, it must be true that $\mu/(1 - \lambda) > 0$. Note that, remarkably, *both* our alternatives are true!

Is $\hat{\mathbf{z}}$ attractive? We can easily evaluate the Jacobian matrix (B2.1) and obtain

$$J = \begin{bmatrix} 1 - \frac{1-\lambda}{\mu} & 1 + \frac{1-\lambda}{\mu} & 0 & \cdots & 0 \\ 0 & 1 & 1 + \frac{1-\lambda}{\mu} & \ddots & \vdots \\ \vdots & \ddots & \ddots & \ddots & 0 \\ \vdots & & 0 & 1 & 1 + \frac{1-\lambda}{\mu} \\ 0 & \cdots & & \cdots & 0 & 1 \end{bmatrix}.$$

The fixed point cannot be attractive! If $(1 - \lambda)/\mu < 2$, it is a *saddle* and attracts a lower-dimensional (actually, a single-dimensional) manifold, otherwise it is a *repellor*.

Other matrix pencils $\{A, B\}$ can be considered, with similar conclusion: fixed points are rare (recall our condition $\mu/(1 - \lambda) > 0$, not very likely for arbitrary complex numbers λ and $\mu \ldots$), and their attractivity is a yet rarer event. This, actually, is why (B1.1) is such a remarkable object to study: Instead of a tedious approach to a limit, the sequence $\{\mathbf{z}_n\}_{n=0}^{\infty}$ displays a much richer dynamics. We have already seen this for the one-dimensional map (B1.2), but a more interesting case will be considered in Section 4. First, however, we need to take a closer look at another intriguing map.

B3. An Intermezzo:
The Map $e_n + 1 = f_n\,|e_n - 1|$

Let $\rho > 0$ be given. We consider the iterative scheme

$$e_{n+1} = f_n|e_n - 1|, \qquad (B3.1)$$

where $\{f_n\}_{n=0}^{\infty}$ is a given positive sequence, uniformly bounded away from both 0 and ∞,

$$0 < f_- := \inf\{f_\ell : \ell \geq 0\} \leq f_n \leq f_+ := \sup\{f_\ell : \ell \geq 0\} < 1.$$

Its importance to our analysis will be made abundantly clear in the next section. The special case of $f_n \equiv \rho \in (0,1)$, $n = 0, 1, \ldots$, deserves a displayed equation on its own,

$$e_{n+1} = \rho|e_n - 1|. \qquad (B3.2)$$

The equation (B3.2) is equivalent (under a simple linear transformation) to the *tent map* – more about it later. . . .

The initial value e_0 in (B3.1) is, in principle, an arbitrary real number. Note, however, that $e_1 > 0$ if $e_0 \neq 1$. Moreover, suppose that $e_n \geq 1$ for some $n \geq 0$. Then $e_{n+1} = f_n(e_n - 1) < e_n - 1$, $e_{n+2} < e_n - 2$ and so on, until we "hit" the interval $(0, 1)$. In other words, after a finite number of steps, m say, we have $e_m \in (0, 1)$. Finally, we take cognisance of the fact that $(0, 1)$ remains *invariant* under the map (B3.1):

$$e_n \in (0, 1) \qquad \Longrightarrow \qquad e_{n+1} \in (0, 1).$$

All this implies that, sooner or later, the sequence $\{e_n\}$ will reach $(0, 1)$ and stay there forever, no matter where the initial value was. Since our interest focuses on the dynamics of this sequence, we may discard any finite number of elements with no damage whatsoever. Consequently, we assume without loss of generality that $e_0 \in (0, 1)$.

Proposition 1.   *Let $\varpi := f_- f_+$. It is true for all $n = 0, 1, \ldots$ that*

(1) $e_{2n} \leq f_+(1 - f_-)\dfrac{1 - \varpi^n}{1 - \varpi} + \varpi^n e_0,$

$$\textbf{(2)} \quad e_{2n} \geq f_-(1-f_+)\frac{1-\varpi^n}{1-\varpi} + \varpi^n e_0,$$

$$\textbf{(3)} \quad e_{2n+1} \leq f_+ - (1-f_+)\varpi\frac{1-\varpi^n}{1-\varpi} - \varpi^n f_+ e_0,$$

$$\textbf{(4)} \quad e_{2n+1} \geq f_- - (1-f_-)\varpi\frac{1-\varpi^n}{1-\varpi} - \varpi^n f_- e_0.$$

Proof. Clearly, both **(1)** and **(2)** are true for $n = 0$. The hypothesis $e_0 \in (0,1)$ implies that

$$e_1 = f_0(1-e_0) \begin{cases} \leq f_+(1-e_0) \\ \geq f_-(1-e_0) \end{cases}$$

and also **(3)** and **(4)** are true for $n = 0$. We continue by induction: Suppose that the proposition is true up to some $n \geq 0$. Since $e_0, \ldots, e_{2n+1} \in (0,1)$, we have, by virtue of **(4)**,

$$e_{2n+2} = f_{2n+1}(1 - e_{2n+1}) \leq f_+(1 - e_{2n+1})$$

$$\leq f_+ \left(1 - f_- + (1-f_-)\varpi\frac{1-\varpi^n}{1-\varpi} + \varpi^n f_- e_0\right)$$

$$= f_+(1-f_-)\frac{1-\varpi^{n+1}}{1-\varpi} + \varpi^{n+1} e_0.$$

This allows us to "advance" the induction a single unit for **(1)**. Likewise, we use **(3)**, **(2)** and **(1)** to "advance" **(2)**, **(3)** and **(4)**, respectively, and the proof follows. □

Corollary. Given $\varepsilon > 0$, there exists N_ε such that for any $e_0 \in \mathcal{R}$ and all $n \geq N_\varepsilon$,

$$\frac{f_- - \varpi}{1-\varpi} - \varepsilon \leq e_n \leq \frac{f_+ - \varpi}{1-\varpi} + \varepsilon.$$

Proof. By letting $n \to \infty$ in **(1)**–**(4)** and since $0 < \varpi < 1$. □

Let us choose $\rho \in (0,1)$ and define positive numbers $\alpha_0, \alpha_1, \ldots$ in the following manner:

$$\alpha_0 := 1,$$

$$\alpha_n := \rho^{-n} f_0 f_1 \cdots f_{n-1}, \quad n = 1, 2, \ldots.$$

Therefore,

$$f_n = \rho \frac{\alpha_{n+1}}{\alpha_n}, \quad n = 0, 1, \ldots.$$

We have already seen that it can be assumed without loss of generality (as far as the dynamics of (B3.1) are concerned) that $e_0 \in (0, 1)$. Then all the e_n's stay in $(0, 1)$, and (B3.1) becomes simply

$$e_{n+1} = f_n(1 - e_n), \quad n = 0, 1, \ldots. \tag{B3.3}$$

It follows at once from (B3.3) by induction that

$$e_n = \sum_{\ell=0}^{n-1} (-1)^\ell \left(\prod_{j=0}^{\ell} f_{n-1-j} \right) + (-1)^n \left(\prod_{j=0}^{n-1} f_j \right) e_0$$

$$= \alpha_n \sum_{\ell=0}^{n-1} \frac{(-1)^\ell}{\alpha_{n-1-\ell}} \rho^{\ell+1} + (-1)^n \rho^n \alpha_n e_0.$$

Let us suppose that the α_n's are uniformly bounded (actually, a weaker assumption will do, namely that $\limsup_{n\to\infty} \rho^n \alpha_n = 0$). Then, for large n, the contribution of e_0 progressively disappears and we can assume without loss of generality that $e_0 = 0$. This simplifies the discussion somewhat and yields

$$e_n = \rho \alpha_n \sum_{\ell=0}^{n-1} \frac{(-1)^\ell}{\alpha_{n-1-\ell}} \rho^\ell. \tag{B3.4}$$

We next assume that there exists an $L_2[-1, 1]$ function g such that

$$\frac{1}{\alpha_n} = g(\cos(\vartheta + n\psi)), \quad n = 0, 1, \ldots,$$

for some numbers ϑ and ψ. Expanding into Fourier series, we have

$$\frac{1}{\alpha_n} = \sum_{k=-\infty}^{\infty} g_k e^{ikn\psi}.$$

Let us substitute the last expression into (B3.4):

$$e_n = \rho \alpha_n \sum_{\ell=0}^{n-1} (-1)^\ell \rho^\ell \sum_{k=-\infty}^{\infty} g_k e^{ik(n-\ell-1)\psi} = \alpha_n \sum_{k=-\infty}^{\infty} g_k d_n(e^{ik\psi}),$$

where

$$d_n(z) = \sum_{\ell=0}^{n-1} (-1)^{n-1-\ell} \rho^{n-\ell} z^\ell = \frac{(-1)^{n-1}\rho^n + z^n}{1 + \rho^{-1}z}.$$

Thus,

$$e_n = \alpha_n \sum_{k=-\infty}^{\infty} g_k \frac{(-1)^{n-1}\rho^n + e^{ikn\psi}}{1 + \rho^{-1}e^{ik\psi}}.$$

Recall that $\rho \in (0,1)$. Thus, for large n,

$$e_n \approx \hat{e}_n := \alpha_n \sum_{k=-\infty}^{\infty} g_k \frac{e^{ikn\psi}}{1 + \rho^{-1}e^{ik\psi}}.$$

In other words, for $n \to \infty$ the point e_n lies arbitrarily close to $\alpha_n \Gamma_n$, where

$$\Gamma_t := \sum_{k=-\infty}^{\infty} g_k \frac{e^{itk\psi}}{1 + \rho^{-1}e^{ik\psi}}.$$

An important consequence is that, in general, $\{e_n\}_{n=0}^{\infty}$ does not tend to a limit. Instead, if ψ/π is rational, the "asymptotic sequence" $\{\hat{e}_n\}_{n=0}^{\infty}$ is periodic. Otherwise it is dense on the interval

$$\left(\inf_{t \geq 0} \Gamma_t, \sup_{t \geq 0} \Gamma_t \right).$$

In the next section, we will take advantage of the aforementioned analysis of (B3.1). We conclude this section by a more substantive set of results on the simplified map (B3.2). As long as ρ lies in $(0,1)$, it is clear that $e_n \to \rho/(1+\rho)$. This we can prove from either the corollary to Proposition 1 (since $f_- = f_+ = \rho$, $\varpi = \rho^2$) or by Fourier analysis (since $\alpha_n \equiv 1$, we have $g(z) \equiv 1$, thus $g_0 = 1$, $g_k = 0$ for $k \neq 0$, and $\Gamma_t \equiv \rho/(1+\rho)$), although a diligent reader might wish to affirm the limit by an elementary method.

The situation is nontrivial already for $\rho = 1$: for any point $\alpha \in [0, 1]$, the pair $\{\alpha, 1-\alpha\}$ forms a 2-periodic sequence (except that, of course, $\alpha = \frac{1}{2}$ gives rise to a fixed point – this is, incidentally, the limiting case of $\rho/(1 + \rho)$ as $\rho \to 1$). Given $e_0 \notin [0, 1]$, we reach the unit interval in a finite number of steps and, subsequently, stay on the orbit $\{\alpha, 1 - \alpha\}$, where α is the fractional part of e_0.

Much can be said also about the behavior of (B3.2) for $\rho > 1$, since it is nothing else than the *tent map*

$$q_{n+1} = 1 - \rho|q_n|,$$

after the change of variables $q_n = 1 - e_n$. We consider it as acting in the interval $[-1, 1]$ to itself – clearly, it is of the form

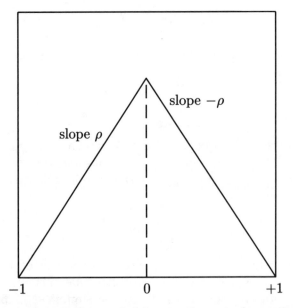

The theory of the tent map has been exhaustively studied and is quite comprehensively understood (cf. [van Strien 1988]). True to our aims in this article, we restrict our discussion to elementary mathematical tools and "back of the envelope" methods.

Having stated this, let us just mention that $\rho > 2$ causes divergence, whereas in the regime $1 < \rho \leq 2$, the map displays quite a convoluted pattern of behavior. A point $x \in [-1, 1]$ is called *nonwandering* if there exists a sequence $\{x_n\}_{n=0}^{\infty} \subset [-1, 1]$ such that $\lim_{n \to \infty} x_n = x$ and such that $\lim_{n \to \infty} q_{m(n)} = x$ (starting with $q_0 = x$) for some subsequence $\{m(n)\}_{n=0}^{\infty}$. The alternative to x being non-wandering is either that it lies in a basin of attraction of a periodic attractor or that it belongs to a *wandering interval* $\mathcal{I}$: Letting $Q(x) = 1 - \rho|x|$ and $Q^{\circ(m)}$ being the mth iterate of Q (that is, a superposition of Q with itself m times), elements of the sequence $\mathcal{I}, Q(\mathcal{I}), Q^{\circ(2)}(\mathcal{I}), \ldots$ are disjoint, and every point of $\mathcal{I}$ is neither eventually periodic or contained in a basin of an attractive periodic orbit. The set of all the nonwandering points in $[0, 1]$ is denoted by Ω.

The set Ω is explicitly known and its structure depends, as one can expect, on the value of $\rho \in (1, 2]$ or, more concretely, on the integer $M = M(\rho)$ such that $\sqrt{2} < \rho^M \leq 2$. The set Ω is a union of M disjoint intervals $\mathcal{I}_0, \mathcal{I}_1, \ldots, \mathcal{I}_{M-1}$ and a finite number of periodic points. Moreover, any member of $\mathcal{I}_\ell$ is mapped into an element of $\mathcal{I}_{(\ell+1) \bmod M}$.

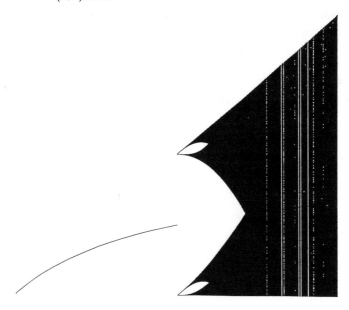

The previous figure displays the *bifurcation diagram* of (B3.2) for $0 < \rho < 2$. The variable ρ varies along the x-axis from 0 to 2. For each such value of ρ, 40000 iterations of (B3.2) were performed, commencing from $e_0 = 1$. The last 20000 iterations are displayed along the y-axis. As expected, we reach a fixed point for $\rho < 1$, hence the diagram consists of the single curve $\rho/(1+\rho)$ there. However, for $\rho > 1$, most values produce "black" intervals, filled by the iterates – these are precisely the sets $\mathcal{I}_\ell$. In the "white" gaps, the iteration is eventually periodic. Everything is as predicted by the theory of the tent map....

Were we to display a bifurcation diagram of (B3.1), with a "sensible" choice of f_n's (and we defer to the next section a discussion of such "sensibility"), the pattern would have been quite similar: tendence to a limit, followed by chaotic behavior with windows of periodicity. Unfortunately, we can no longer use the relatively simple theory of the tent map to underpin this observation.

B4. A Simplified Map

We can add with complete impunity an extra scalar unknown to the map (B1.1). Thus, let

$$\mathbf{w}_{n+1}^{(\alpha)} = A\mathbf{w}_n^{(\alpha)} + \frac{\alpha}{\|\mathbf{w}_n^{(\alpha)}\|} B\mathbf{w}_n^{(\alpha)}, \qquad (\text{B4.1})$$

where $\alpha \in \mathcal{C} \setminus \{0\}$. Letting

$$\mathbf{z}_n := \frac{1}{\alpha}\mathbf{w}_n^{(\alpha)}, \quad n = 0, 1, \ldots,$$

it follows readily that

$$\mathbf{z}_{n+1} = A\mathbf{z}_n + \frac{e^{i\omega}}{\|\mathbf{z}_n\|} B\mathbf{z}_n,$$

where $e^{i\omega} = \alpha/|\alpha|$. This is exactly the map (B1.1), except that the matrix B might need to be rotated. Hence, whatever we might find for (B1.1) can be suitably adjusted to cater for (B4.1).

In the present section, we restrict our attention to the special Theodorus spiral

$$\mathbf{w}_{n+1}^{(\alpha)} = \left(1 - \frac{\alpha}{\|\mathbf{w}_n^{(\alpha)}\|}\right) A\mathbf{w}_n^{(\alpha)}, \tag{B4.2}$$

where $\alpha > 0$. By virtue of the analysis in the previous paragraph, $\omega = 0$ and (B4.2) transforms into

$$\mathbf{z}_{n+1} = \left(1 - \frac{1}{\|\mathbf{z}_n\|}\right) A\mathbf{z}_n. \tag{B4.3}$$

In the present section, we analyze the behavior of (B4.3).

We stipulate that A possesses a full set of eigenvectors, $\mathbf{v}_1, \ldots,$ $\mathbf{v}_d$, say. The corresponding eigenvalues will be denoted by $\lambda_1, \ldots, \lambda_d$. Since the eigenvectors span $\mathcal{R}^d$, we can express $\mathbf{z}_n$ in that basis,

$$\mathbf{z}_n = \sum_{\ell=1}^{d} \alpha_\ell^{(n)} \mathbf{v}_\ell, \quad n = 0, 1, \ldots.$$

It follows from (B4.3) that

$$\mathbf{z}_{n+1} = \left(1 - \frac{1}{\|\mathbf{z}_n\|}\right) \sum_{\ell=1}^{d} \alpha_\ell^{(n)} \lambda_\ell \mathbf{v}_\ell,$$

therefore

$$\alpha_\ell^{(n+1)} = \left(1 - \frac{1}{\|\mathbf{z}_n\|}\right) \lambda_\ell \alpha_\ell^{(n)}, \quad n = 0, 1, \ldots. \tag{B4.4}$$

It now follows readily by induction that

$$\alpha_\ell^{(n)} = \left\{\prod_{j=0}^{n-1} \left(1 - \frac{1}{\|\mathbf{z}_j\|}\right)\right\} \lambda_\ell^n \alpha_\ell^{(0)}. \tag{B4.5}$$

Let us stipulate that the eigenvalues are ordered so that

$$|\lambda_1| \geq |\lambda_2| \geq \cdots \geq |\lambda_d|,$$

and denote by ℓ^* the least index such that $\alpha_{\ell^*}^{(0)} \neq 0$. There exists some $k^* \in \{\ell^*, \ldots, d\}$ such that

$$|\lambda_{\ell^*}| = \cdots = |\lambda_{k^*}| > |\lambda_{k^*+1}|.$$

It follows from (B4.5) that, for large n,

$$\mathbf{z}_n \approx \left\{ \prod_{j=0}^{n-1} \left(1 - \frac{1}{\|\mathbf{z}_j\|} \right) \right\} \sum_{\ell=\ell^*}^{k^*} \lambda_\ell^n \alpha_\ell^{(0)} \mathbf{v}_\ell.$$

In other words, $\mathbf{z}_n$ approaches the $(k^*-\ell^*+1)$-dimensional linear subspace that is spanned by the eigenvectors $\mathbf{v}_{\ell^*}, \ldots, \mathbf{v}_{k^*}$.

Henceforth we assume that $\ell^* = 1$ – the general case follows in a very similar manner, albeit with considerably more unpleasant notation. Thus, k^* is the dimension of the invariant subspace where the whole "action" takes place. The simplest case is, obviously, $k^* = 1$: now

$$\mathbf{z}_n \approx \alpha_1^{(0)} \prod_{j=0}^{n-1} \left(1 - \frac{1}{\|\mathbf{z}_j\|} \right) \lambda_1^n \mathbf{v}_1 = \alpha_1^{(n)} \mathbf{v}_1.$$

In particular, by (B4.4) and letting $\|\mathbf{v}_1\| = 1$,

$$\|\mathbf{z}_{n+1}\| \approx \left| \alpha_1^{(n+1)} \right| = \left| 1 - \frac{1}{\|\mathbf{z}_n\|} \right| |\lambda_1| \left| \alpha_1^{(n)} \right| \approx |\lambda_1| \left| 1 - \|\mathbf{z}_n\| \right|.$$

In other words, the norms $\{\|\mathbf{z}_n\|\}$ behave asymptotically like the solutions $\{e_n\}$ of the one-dimensional map (B3.2) with $\rho = |\lambda_1|$.

Let

$$\frac{\alpha_1^{(0)}}{|\alpha_1^{(0)}|} = e^{i\phi}, \quad \frac{\lambda_1}{|\lambda_1|} = e^{i\tau}, \quad -\pi < \phi, \tau \leq \pi.$$

Then

$$\mathbf{z}_n = e^{i(\phi+n\tau)} \left| \alpha_1^{(0)} \right| \prod_{\ell=0}^{n-1} \left| 1 - \frac{1}{\|\mathbf{z}_\ell\|} \right| \mathbf{v}_1,$$

consequently

$$\mathbf{z}_n = e^{i(\phi+n\tau)} \|\mathbf{z}_n\| \mathbf{v}_1.$$

Clearly, even if $\|\mathbf{z}_n\|$ approaches a limit (that is, as long as $|\lambda_1| < 1$), the "angle" of $\mathbf{z}_n$ goes on changing indefinitely, unless $\tau = 0$. Moreover, in line with the results on the tent map, as reported in the previous section, the behavior of $\{\mathbf{z}\}_{n=0}^{\infty}$ is either periodic or chaotic for $\rho \in [1, 2]$. Note that we have two mechanisms that produce "interesting" behavior, namely the modulus $\|\mathbf{z}_n\|$ (which might be periodic or chaotic) via the angle τ (which is either periodic or ergodic).

As is only fair to expect, $k^* \geq 2$ brings about considerably more complicated behavior. However, is the case of "nontrivial" k^* really interesting? After all, how likely is it for several eigenvalues to dominate in unison? Quite likely, actually! Multiple dominant eigenvalues extensively occur, for example, in the Frobenius–Perron theory of matrices with positive entries [Varga 1962]. More importantly, $k^* = 2$ is bound to occur whenever the dominant eigenvalues of a *real* matrix A appear as a complex conjugate pair. This is quite a common event, and we devote the remainder of this section to its analysis.

Since the "action" takes place in a two-dimensional subspace, we might assume just as well that $d = 2$. Thus, let $\lambda = \lambda_1$, therefore $\lambda_2 = \bar{\lambda}$. Clearly, $\text{Im}\,\lambda \neq 0$ is necessary for the eigenvalues to form a genuinely complex conjugate pair, but then $\lambda \neq \bar{\lambda}$ and A possesses a spectral factorization

$$A = V\Lambda V^{-1}, \qquad V = \begin{bmatrix} v_1 & \bar{v}_1 \\ v_2 & \bar{v}_2 \end{bmatrix}, \qquad \Lambda = \begin{bmatrix} \lambda & 0 \\ 0 & \bar{\lambda} \end{bmatrix}.$$

The eigenvectors

$$\begin{bmatrix} v_1 \\ v_2 \end{bmatrix} \quad \text{and} \quad \begin{bmatrix} \bar{v}_1 \\ \bar{v}_2 \end{bmatrix}$$

can be normalized arbitrarily, but it is convenient to require that

$$|v_1|^2 + |v_2|^2 = \tfrac{1}{2}.$$

Let $\mathbf{w}_n := V^{-1}\mathbf{z}_n$, $n = 0, 1, \ldots$. Thus, (B4.3) yields

$$\mathbf{w}_{n+1} = \left(1 - \frac{1}{\|V\mathbf{w}_n\|}\right)\Lambda\mathbf{w}_n, \quad n = 0, 1, \ldots, \tag{B4.6}$$

where

$$\mathbf{w}_0 = V^{-1}\mathbf{z}_0 = \frac{1}{2\,\mathrm{Im}\,v_1\bar{v}_2}\left\{\begin{bmatrix}\mathrm{Im}\,(v_1 z_{0,2} - v_2 z_{0,1})\\ \mathrm{Im}\,(v_1 z_{0,2} - v_2 z_{0,1})\end{bmatrix}\right.$$

$$\left.+i\begin{bmatrix}\mathrm{Re}\,(v_1 z_{0,2} - v_2 z_{0,1})\\ \mathrm{Re}\,(-v_1 z_{0,2} + v_2 z_{0,1})\end{bmatrix}\right\} := \begin{bmatrix}\omega_0\\ \bar{\omega}_0\end{bmatrix}.$$

It follows easily by induction in (B4.6) that

$$\mathbf{w}_n = \begin{bmatrix}\omega_n\\ \bar{\omega}_n\end{bmatrix}, \quad n = 0, 1, \dots,$$

where the ω_n's obey the recurrence

$$\omega_{n+1} = \left(1 - \frac{1}{\|V\mathbf{w}_n\|}\right)\lambda\omega_n, \quad n = 0, 1, \dots. \tag{B4.7}$$

Thus, (B4.6) can be expressed as a one-dimensional map, except that we must somehow get rid of the term $\|V\mathbf{w}_n\|$ in the denominator. Fortunately,

$$\|V\mathbf{w}_n\|^2 = |v_1\omega_n + \bar{v}_1\bar{\omega}_n|^2 + |v_2\omega_n + \bar{v}_2\bar{\omega}_n|^2$$

$$= |\omega_n|^2 + 2\mathrm{Re}\,(v_1^2 + v_2^2)\omega_n^2,$$

since $|v_1|^2 + |v_2|^2 = \frac{1}{2}$. This allows us to express all of (B4.7) in the single currency of ω_n's:

$$\omega_{n+1} = \left(1 - \frac{1}{\{|\omega_n|^2 + 2\mathrm{Re}\,((v_1^2 + v_2^2)\omega_n^2)\}^{1/2}}\right)\lambda_n\omega_n.$$

Let

$$\omega_n = r_n e^{i\theta_n}, \quad \lambda = \rho e^{\frac{1}{2}i\psi}, \quad v_1^2 + v_2^2 = \sigma e^{i\varphi}, \quad \vartheta := \varphi + 2\theta_0,$$

where $r_n, \rho \geq 0$ and $\theta_n, \psi, \varphi \in (-\pi, \pi]$. Note that $|\sigma| < 1$, because of $|v_1|^2 + |v_2|^2 = \frac{1}{2}$. More importantly, it is obvious from

(B4.7) that $\theta_n = \theta_0 + \frac{1}{2}n\psi$, $n = 0, 1, \ldots$. Again, the orientation can be resolved easily! We now have

$$\|V\mathbf{w}_n\|^2 = r_n^2 \left(1 + 2\mathrm{Re}\left((v_1^2 + v_2^2)e^{i(2\theta_0 + n\psi)}\right)\right)$$
$$= r_n^2(1 + \sigma\cos(\vartheta + 2n\psi))$$

and obtain from (B4.7) a map for the moduli $\{r_n\}$, namely

$$r_{n+1} = |\lambda| \left| r_n - \frac{1}{\sqrt{1 + \sigma\cos(\vartheta + n\psi)}} \right|, \quad n = 0, 1, \ldots. \quad \text{(B4.8)}$$

Letting

$$e_n = r_n\sqrt{1 + \sigma\cos(\vartheta + n\psi)}, \quad n = 0, 1, \ldots,$$

we can recast (B4.8) in the form

$$e_{n+1} = |\lambda| \left\{ \frac{1 + \sigma\cos(\vartheta + (n+1)\psi)}{1 + \sigma\cos(\vartheta + n\psi)} \right\}^{1/2} |e_n - 1|, \quad n = 0, 1, \ldots.$$
$$\text{(B4.9)}$$

It is now obvious why we have paid all this attention to the map (B3.1), since (B4.9) is precisely in that form, with

$$f_n = |\lambda| \left\{ \frac{1 + \sigma\cos(\vartheta + (n+1)\psi)}{1 + \sigma\cos(\vartheta + n\psi)} \right\}^{1/2}.$$

In particular,

$$f_- = |\lambda|\sqrt{\frac{1 - |\sigma|}{1 + |\sigma|}}, \quad f_+ = |\lambda|\sqrt{\frac{1 + |\sigma|}{1 - |\sigma|}}$$

(recall that $|\sigma| < 1$). It follows from the corollary to Proposition 1 that, as long as $\varpi = |\lambda|^2 < 1$, it is true that for every $\varepsilon > 0$ and sufficiently large n,

$$|\lambda|\frac{\sqrt{\frac{1 - |\sigma|}{1 + |\sigma|}} - |\lambda|}{1 - |\lambda|^2} - \varepsilon \leq e_n \leq |\lambda|\frac{\sqrt{\frac{1 + |\sigma|}{1 - |\sigma|}} - |\lambda|}{1 - |\lambda|^2} + \varepsilon$$

(strictly speaking, the corollary was proved for $f_+ < 1$, but it is quite clear that the proof remains perfectly valid for $\varpi < 1$).

Thus, as long as the eigenvalue λ is in the unit disk, the sequence $\{e_n\}_{n=0}^{\infty}$ is uniformly bounded. Moreover, if $|\lambda| < ((1-|\sigma|)/(1+|\sigma|))^{1/2}$, then, in addition, it is bounded away from zero and, asymptotically, its values are confined to a positive interval.

Next we observe that, in the notation of Section 3,

$$\alpha_n = \sqrt{1 + \sigma\cos(\vartheta + n\psi)}, \quad n = 0, 1, \ldots,$$

whereas ρ plays here the same role as in that section. Thus,

$$g(\cos(\vartheta + n\psi)) = \frac{1}{\sqrt{1 + \sigma\cos(\vartheta + n\psi)}},$$

hence

$$g(z) = \frac{1}{(1 + \sigma z)^{1/2}} = \sum_{m=0}^{\infty} (-1)^m \frac{(2m)!}{(m!)^2} \left(\frac{\sigma z}{4}\right)^m.$$

Recall that we are interested in the Fourier coefficients of $g(\cos(\vartheta + n\psi))$. They are given explicitly by

$$g_k = \frac{1}{2\pi} \int_{-\pi}^{\pi} e^{-ik\tau} g(\cos(\vartheta + \tau)) d\tau$$

$$= \frac{1}{2\pi} \int_{-\pi}^{\pi} e^{-ik\tau} \sum_{m=0}^{\infty} \frac{(2m)!}{(m!)^2} \left(-\frac{\sigma}{4}\right)^m \cos^m(\vartheta + \tau) d\tau$$

$$= \sum_{m=0}^{\infty} \frac{(2m)!}{(m!)^2} \left(-\frac{\sigma}{4}\right)^m h_k^{(m)},$$

where

$$h_k^{(m)} := \frac{1}{2\pi} \int_{-\pi}^{\pi} e^{-ik\tau} \cos^m(\vartheta + \tau) d\tau, \quad m = 0, 1, \ldots, \; k \in \mathcal{Z}.$$

Since

$$\cos^m \gamma = \tfrac{1}{2}(e^{i\gamma} + e^{-i\gamma}) \cos^{m-1} \gamma,$$

we derive the recurrence

$$h_k^{(m)} = \tfrac{1}{2}\left(e^{i\vartheta}h_{k-1}^{(m-1)} + e^{-i\vartheta}h_{k+1}^{(m-1)}\right), \quad m = 1, 2, \dots, \; k \in \mathcal{Z}.$$

But

$$h_0^{(0)} = 1, \quad h_k^{(0)} = 0, \; k \in \mathcal{Z} \setminus \{0\},$$

and it follows readily by induction that

$$h_{-m+2\ell}^{(m)} = \frac{1}{2^m}\binom{m}{\ell}e^{i(2\ell-m)\vartheta}, \quad \ell = 0, 1, \dots, m,$$

and $h_k^{(m)} = 0$ for all $|k| \geq m+1$. The Fourier coefficients can now be written concisely in terms of *hypergeometric functions* [Rainville 1967]: given any three complex numbers a, b, c, where c is neither zero nor a negative integer, the hypergeometric function is defined as

$$F\begin{bmatrix} a, b; \\ c; \end{bmatrix} z \end{bmatrix} := \sum_{m=0}^{\infty} \frac{(a)_m (b)_m}{(c)_m} \frac{z^m}{m!},$$

where

$$(x)_0 :\equiv 1,$$

$$(x)_m := x(x+1)(x+2)\cdots(x+m-1)$$

$$= (x)_{m-1}(x+m-1), \quad m = 1, 2, \dots,$$

is the *Pochhammer symbol*, also known as the *generalized factorial*. Note that, as long as x is neither zero nor a negative integer, $(x)_m = \Gamma(x+m)/\Gamma(x)$, where Γ is the familiar gamma function.

Since

$$\frac{(2m)!}{4^m m!} = \left(\frac{1}{2}\right)_m, \quad m = 0, 1, \dots,$$

we have, after long, tedious but perfectly straightforward calculation,

$$g_{2k} = \sum_{m=0}^{\infty} \frac{\left(\frac{1}{2}\right)_m}{m!} (-\sigma)^m h_{2k}^{(m)}$$

$$= e^{2ki\vartheta} \sum_{m=|k|}^{\infty} \frac{\left(\frac{1}{2}\right)_{2m}}{(2m)!} \left(\frac{\sigma}{2}\right)^{2m} \binom{2m}{m+|k|}$$

$$= e^{2ki\vartheta} \sum_{m=0}^{\infty} \frac{\left(\frac{1}{2}\right)_{2m+2|k|}}{m!(m+2|k|)!} \left(\frac{\sigma}{2}\right)^{2(m+|k|)}$$

$$= e^{2ki\vartheta} \sum_{m=0}^{\infty} \frac{\left(\frac{1}{4}\right)_{m+|k|} \left(\frac{3}{4}\right)_{m+|k|}}{m!(m+2|k|)!} \sigma^{2(m+|k|)}$$

$$= e^{2ki\vartheta} \frac{\left(\frac{1}{4}\right)_{|k|} \left(\frac{3}{4}\right)_{|k|}}{(2|k|)!} \sigma^{2|k|} F\left[\begin{matrix} |k|+\frac{1}{4}, |k|+\frac{3}{4}; \\ 2|k|+1; \end{matrix} \sigma^2\right]$$

for all integer k. Likewise, we obtain

$$g_{2k+1} = -e^{(2k+1)i\vartheta} \frac{\left(\frac{1}{4}\right)_{|k|+1} \left(\frac{3}{4}\right)_{|k|}}{(2|k|+1)!} \sigma^{2|k|+1}$$

$$\times F\left[\begin{matrix} |k|+\frac{3}{4}, |k|+\frac{5}{4}; \\ 2|k|+2; \end{matrix} \sigma^2\right].$$

Recall from Section 3 that, for every $\varepsilon > 0$, there exists N_ε so that $n \geq N_\varepsilon$ implies that $|e_n - \hat{e}_n| < \varepsilon$, where

$$\hat{e}_n = \alpha_n \sum_{k=-\infty}^{\infty} g_k \frac{e^{ikn\psi}}{1 + \rho^{-1} e^{ik\psi}}.$$

Moreover, $r_n = e_n/\alpha_n$, thus, asymptotically, the numbers r_n are arbitrarily close to

$$
\hat{r}_n = \sum_{k=-\infty}^{\infty} \frac{(-\sigma)^{|k|} \left(\frac{1}{4}\right)_{[(|k|+1)/2]} \left(\frac{3}{4}\right)_{[|k|/2]}}{|k|!(1 + \rho^{-1} e^{ik\psi})}
$$

$$
\times F\left[\begin{matrix} [(|k|+1)/2] + \frac{1}{4}, [|k|/2] + \frac{3}{4}; \\ |k| + 1; \end{matrix} \sigma^2\right] e^{ik(\vartheta + n\psi)}
$$

$$
= \frac{1}{2} p_0 \frac{1}{1 + \rho^{-1}} + \sum_{k=1}^{\infty} p_k \frac{\cos(\vartheta + n\psi) + \rho^{-1} \cos\vartheta}{1 + 2\rho^{-1} \cos k\psi + \rho^{-2}},
$$

where

$$
p_k := 2(-\sigma)^k \left(\frac{1}{4}\right)_{[(k+1)/2]} \left(\frac{3}{4}\right)_{[k/2]}
$$

$$
\times F\left[\begin{matrix} [(k+1)/2] + \frac{1}{4}, [k/2] + \frac{3}{4}; \\ k + 1; \end{matrix} \sigma^2\right], \quad k = 0, 1, \ldots.
$$

In other words, the r_n's lie asymptotically on the interval

$$
\left\{\frac{1}{2} p_0 \frac{1}{1 + \rho^{-1}} + \sum_{k=1}^{\infty} p_k \frac{\cos(\vartheta + t\psi) + \rho^{-1} \cos\vartheta}{1 + 2\rho^{-1} \cos k\psi + \rho^{-2}} : t \geq 0\right\},
$$

at integer values of t. True to the pattern that has been already established for "simpler" cases, this might lead either to accumulation at a finite set of points (an eventually-periodic sequence) if ψ/π is rational or to a space-filling behavior otherwise.

We now travel all the way back from the r_n's to the z_n's: we have

$$
\omega_n = r_n e^{i(\theta_0 + \frac{1}{2} n\psi)}
$$

and

$$
z_n = 2\mathrm{Re}\begin{bmatrix} v_1 \omega_n \\ v_2 \omega_n \end{bmatrix}.
$$

Hence, we can write the attractor of the sequence $z_0, z_1, \ldots$ explicitly, although it is not very illuminating. However, we can conclude – and this provides a great deal of insight – that in the case $|\lambda| < 1$, the z_n's lie asymptotically on a limit cycle in $\mathcal{C}^2$ and discern the pattern of their behavior there: they can either tend to a finite periodic set or fill the whole cycle.

Our point of departure was the map (B4.2) with $\alpha > 0$. We may just as well contemplate that map with $\alpha < 0$. This gives

$$z_{n+1} = \left(1 + \frac{1}{\|z_n\|}\right) A z_n. \qquad (B4.10)$$

The dynamics of (B4.10) can be worked out similarly to these of (B4.3). Again, the most interesting case is $d = 2$, with a complex conjugate pair of eigenvalues. Maintaining an identical notation – and sparing the reader all the details – we can prove that, as long as $|\lambda| < 1$, $r_n \approx \Gamma_n$ for $n \gg 1$, where

$$\Gamma_t = \sum_{k=-\infty}^{\infty} \frac{g_k e^{ikt\psi}}{-1 + \rho^{-1} e^{ik\psi}}, \quad t \geq 0.$$

Figures 40–43 display the attractor of (B4.2) for

$$A = a \begin{bmatrix} 1.5 & 0.71 \\ -0.41 & 0.58 \end{bmatrix},$$

where a is 0.7, 0.95, 1 and 1.25 respectively. We "travel" through three regimes: In Figure 40 the iterants sort themselves out on a limit cycle that becomes a "butterfly" in Figure 41. As a (hence ρ) increases, so does the chaos. This gives the "palette" in Figure 42 and, finally, the "spooky palette" in Figure 43. Similar patterns persist for a wide variety of examples. Thus, in Figures 44 and 45, we display the attractors of

$$A = a \begin{bmatrix} 1.1 & 0.71 \\ -0.65 & 0.58 \end{bmatrix}$$

with $a = 1$ and $a = 1.15$, respectively. Figure 46 originates in

$$A = \begin{bmatrix} 1.0465 & 0.8165 \\ -0.7475 & 0.6670 \end{bmatrix}.$$

Finally, Figure 47 displays the "evolution" of

$$A = \begin{bmatrix} \frac{1}{2} - b & b \\ -2b & \frac{1}{2} + b \end{bmatrix}$$

with four choices of the parameter, namely $b = \frac{1}{3}, \frac{5}{6}, 1$ and $\frac{4}{3}$, respectively.

B5. Unitary Matrices

Let both A and B be $d \times d$ *unitary* matrices (orthogonal matrices being a special case), such that $A^*B + B^*A = O$. Thus, in (B1.1),

$$\|z_{n+1}\|^2 = \left\| Az_n + B\frac{z_n}{\|z_n\|} \right\|^2$$

$$= \left(Az_n + B\frac{z_n}{\|z_n\|} \right)^* \left(Az_n + B\frac{z_n}{\|z_n\|} \right)$$

$$= z_n^* A^* Az_n + \frac{1}{\|z_n\|} z_n^* (A^*B + B^*A)z_n$$

$$+ \frac{1}{\|z_n\|^2} z_n^* B^* Bz_n = \|z_n\|^2 + 1.$$

Consequently, by induction on n,

$$\|z_n\| = \sqrt{\|z_0\|^2 + n}. \tag{B5.1}$$

We assume for simplicity that $\|z_0\| = 1$ and let $u_n := z_n/\|z_n\|$, $n = 0, 1, \ldots$. Then, with the help of (B5.1), (B1.1) becomes

$$u_{n+1} = \frac{1}{\sqrt{n+2}} \left(\sqrt{n+1} A + B \right) u_n. \tag{B5.2}$$

Note that all the $\mathbf{u}_n$'s "live" on the $(d-1)$-dimensional surface of the sphere $\|\mathbf{u}\| = 1$.

The matrix $W := AB^*$ plays a central role in the analysis of (B5.2). Note that W is itself a unitary matrix and that

$$O = A^*B + B^*A = A^*W^*A + A^*WA = A^*(W + W^*)A. \quad \text{(B5.3)}$$

Let us suppose that $W + W^* \neq O$. Then it has a nonzero eigenvalue: There exist $\lambda \neq 0$ and $\mathbf{v} \in \mathcal{R}^d$, $\mathbf{v} \neq \mathbf{0}$, such that $(W + W^*)\mathbf{v} = \lambda\mathbf{v}$. Set $\mathbf{w} := A^*\mathbf{v}$. Then

$$\mathbf{w}^*A^*(W + W^*)A\mathbf{w} = \mathbf{v}^*(W + W^*)\mathbf{v} = \lambda\|\mathbf{v}\|^2 \neq 0,$$

in contradiction to (B5.3). We conclude that $W + W^* = O$ or, by virtue of unitarity,

$$W^2 + I = O. \quad \text{(B5.4)}$$

In a sense, W is a "generalization" of the pure imaginary number i.

Given $d = 2$, an arbitrary unitary matrix can be written as

$$\begin{bmatrix} e^{i\alpha_1} & 0 \\ 0 & e^{i\alpha_2} \end{bmatrix} \begin{bmatrix} \cos\tau & \sin\tau \\ -\sin\tau & \cos\tau \end{bmatrix} \begin{bmatrix} e^{i\beta_1} & 0 \\ 0 & e^{i\beta_2} \end{bmatrix},$$

where $\tau, \alpha_1, \alpha_2, \beta_1, \beta_2 \in (-\pi, \pi]$. Imposition of (B5.4) is tantamount with the following three equations:

$$0 = \cos(\alpha_1 + \beta_1)\cos\tau; \quad \text{(B5.5)}$$

$$0 = \left(e^{i(\alpha_1 + \beta_2)} + e^{-i(\alpha_2 + \beta_1)}\right)\sin\tau; \quad \text{(B5.6)}$$

$$0 = \cos(\alpha_2 + \beta_2)\cos\tau. \quad \text{(B5.7)}$$

Let $\tau \notin \{0, \pm\frac{\pi}{2}, \pi\}$. Then $\cos\tau, \sin\tau \neq 0$, (B5.5) implies that $\alpha_1 + \beta_1 = \pm\frac{\pi}{2}$, (B5.7) implies that $\alpha_2 + \beta_2 = \pm\frac{\pi}{2}$, whereas, as a consequence of (B5.6), $\alpha_1 + \beta_2 = \alpha_2 + \beta_1 = \pm\frac{\pi}{2}$. It follows that

$$W = \begin{bmatrix} \pm i\cos\tau & \sin\tau \\ -\sin\tau & \mp i\cos\tau \end{bmatrix}.$$

This remains valid when $\tau = \pm\frac{\pi}{2}$, while if $\tau \in \{0, \pi\}$, the remaining case, we have

$$W = \begin{bmatrix} 0 & e^{i\omega} \\ -e^{-i\omega} & 0 \end{bmatrix}$$

for some $\omega \in (-\pi, \pi]$.

We do not require the aforementioned analysis to find all the real (i.e., orthogonal) matrices A and B such that $A^T B + B^T A = O$: They are

$$A = \begin{bmatrix} \cos a & \sin a \\ -\sin a & \cos a \end{bmatrix}, \quad B = \begin{bmatrix} \cos b & \sin b \\ -\sin b & \cos b \end{bmatrix},$$

where $b = a \pm \frac{\pi}{2}$.

Let us first assume that $b = a + \frac{\pi}{2}$. Then $\mathbf{u}_{n+1} = C_n \mathbf{u}_n$, where

$$C_n = \frac{1}{\sqrt{n+2}} \begin{bmatrix} \sqrt{n+1}\cos a - \sin a & \sqrt{n+1}\sin a + \cos a \\ -\sqrt{n+1}\sin a - \cos a & \sqrt{n+1}\cos a - \sin a \end{bmatrix}$$

$$= \frac{1}{2\sqrt{n+1}} \begin{bmatrix} 1 & 1 \\ i & -i \end{bmatrix} \begin{bmatrix} \left(\sqrt{n+1}+i\right)e^{ia} & 0 \\ 0 & \left(\sqrt{n+1}-i\right)e^{-ia} \end{bmatrix}$$

$$\times \begin{bmatrix} 1 & -i \\ 1 & i \end{bmatrix}.$$

All the matrices C_n commute (since they share the same eigenvectors) and

$$\mathbf{u}_n = \frac{1}{2} \begin{bmatrix} 1 & 1 \\ i & -i \end{bmatrix} \begin{bmatrix} w_n & 0 \\ 0 & \bar{w}_n \end{bmatrix} \begin{bmatrix} 1 & -i \\ 1 & i \end{bmatrix},$$

where

$$w_n := e^{ina} \prod_{\ell=0}^{n-1} \frac{\sqrt{\ell+1}+i}{\sqrt{\ell+2}}, \quad n = 0, 1, \ldots.$$

However, recall that $\|\mathbf{u}_n\| = 1$. Consequently, there exists $|\phi| \leq \pi$ such that

$$\mathbf{u}_0 = \begin{bmatrix} \cos \phi \\ \sin \phi \end{bmatrix},$$

and it follows at once that

$$\mathbf{u}_n = \begin{bmatrix} \cos(\phi - na - \chi_n) \\ \sin(\phi - na - \chi_n) \end{bmatrix}, \quad n = 0, 1, \ldots, \tag{B5.8}$$

where

$$e^{i\chi_n} = \frac{\sqrt{\ell+1} + i}{\sqrt{\ell+2}}.$$

Therefore, up to an integer multiple of 2π,

$$\chi_n = \sum_{\ell=1}^{n} \tan^{-1} \frac{1}{\sqrt{\ell}}, \quad n = 1, 2, \ldots. \tag{B5.9}$$

The case $b = a - \frac{\pi}{2}$ is virtually identical and it leads to (B5.8), except that the $-\chi_n$ need to be replaced by $+\chi_n$ throughout.

Recall from the previous sections the two components determining the asymptotic behavior of (B1.1): namely, the moduli and the angles. In the present case, the moduli are fixed and we wish to investigate the asymptotics of the angles $\phi - na \pm \chi_n$ for large n. First we consider the distribution of the values of $\chi_n \bmod 2\pi$ in the interval $[0, 2\pi]$ (rather than in $[-\pi, \pi]$, our practice elsewhere in this article – this leads to somewhat easier expressions).

Let f be a continuously differentiable function for all $x \geq 1$. The celebrated *Euler's summation formula* [Rainville 1967] reads

$$\sum_{\ell=1}^{n} f(\ell) = \int_1^n f(x)\mathrm{d}x + \frac{f(1) + f(n)}{2}$$

$$+ \int_1^n \left(x - [x] - \frac{1}{2} \right) f'(x)\mathrm{d}x.$$

Letting $f(x) = \tan^{-1}(1/\sqrt{x})$ yields

$$\sum_{\ell=1}^{n} \tan^{-1}\frac{1}{\sqrt{\ell}} = \int_{1}^{n} \tan^{-1}\frac{1}{\sqrt{x}}dx + \frac{\pi}{8} + \frac{1}{2}\tan^{-1}\frac{1}{\sqrt{n}}$$

$$+ \int_{1}^{n}\frac{x - [x] - \frac{1}{2}}{\sqrt{x}(1+x)}dx. \qquad (B5.10)$$

But

$$\int_{1}^{y}\tan^{-1}\frac{1}{\sqrt{x}}dx = \int_{1}^{y}\sum_{k=0}^{\infty}\frac{(-1)^k}{2k+1}x^{-k-\frac{1}{2}}dx = G(1) - \sqrt{y}G\left(\frac{1}{y}\right),$$

where

$$G(y) = 2\sum_{k=0}^{\infty}\frac{(-1)^k}{4k^2-1}y^k = -1 - \frac{1+y}{\sqrt{y}}\tan^{-1}\sqrt{y}.$$

Therefore,

$$\int_{1}^{n}\tan^{-1}\frac{1}{\sqrt{t}}dt = \sqrt{n} - 1 - \frac{\pi}{2} + (1+n)\tan^{-1}\frac{1}{\sqrt{n}}.$$

Moreover,

$$\int_{\ell-1}^{\ell}\frac{x-\ell+\frac{1}{2}}{\sqrt{x}(1+x)}dx$$

$$= \int_{-\frac{1}{2}}^{1/2}\frac{x\,dx}{\{x+\ell-\frac{1}{2}\}^{1/2}(x+\ell+\frac{1}{2})}$$

$$= -\int_{0}^{1/2}\frac{\{\ell-\frac{1}{2}+x\}^{1/2}(\ell+\frac{1}{2}+x) - \{\ell-\frac{1}{2}-x\}^{1/2}(\ell+\frac{1}{2}-x)}{\{(\ell-\frac{1}{2})^2 - x^2\}^{1/2}\left((\ell+\frac{1}{2})^2 - x^2\right)}x\,dx$$

$$= 0 - \frac{1}{4\ell^{5/2}} + \mathcal{O}(\ell^{-7/2}).$$

It follows that

$$\int_{1}^{n}\frac{x-[x]-\frac{1}{2}}{\sqrt{x}(1+x)}dx = c_1 + o(1) \overset{n\to\infty}{\longrightarrow} c_1,$$

where c_1 is a finite constant. Substituting all this into (B5.10) shows that $\chi_n = 2\sqrt{n} + c_2 + o(1)$ for all $n \gg 1$, where c_2 is yet another constant.

Recall that our goal is to investigate the behavior of $\phi - na \pm \chi_n$ for $n \gg 1$. Note first that we may consider $\chi_n - na$, since neither uniform rotations nor orientation matter to the qualitative picture. Let firstly $a = 0$. Wishing to prove, as we do, that the sequence $\{\sqrt{n} \bmod 2\pi\}$ is equidistributed in $[0, 2\pi]$ (that is, that the proportion of times it "hits" any interval – or, in general, any measurable subset of $[0, 2\pi]$ – of length L is asymptotically $L/2\pi$), we have several options: We can use the Weyl equidistribution theorem [Körner 1988], the mean ergodic theorem of von Neumann [Halmos 1956] or a theorem of Fejér [Pólya and Szegő 1979, Vol. 1, Pt. II, Problem 175]. Faced with this abundance of riches, we opt instead for a direct proof which contributes an extra useful morsel of information.

Proposition 2.   *The sequence* $\{\sqrt{n} \bmod P\}_{n=0}^{\infty}$, *where* $P > 0$, *is equidistributed in the interval* $[0, P]$.

Proof. Let M be a large integer. We examine all the integers in the interval $[M^2 P^2, (M + 1)^2 P^2)$. Clearly, they are of the form $[M^2 P^2] + k$, $k = 0, 1, \ldots, [(2M + 1)P^2] - 1$. Let $\Omega = \{M^2 P^2\} = M^2 P^2 - [M^2 P^2]$, the fractional part of $M^2 P^2$. Since

$$\sqrt{M^2 P^2 + k - \Omega} = MP \left\{ 1 + \frac{k - \Omega}{M^2 P^2} \right\}^{1/2}$$

$$= MP + \frac{k - \Omega}{2MP} + \mathcal{O}(M^{-2}),$$

we have

$$g_n := \sqrt{M^2 P^2 + k - \Omega} \bmod P = \frac{k - \Omega}{2MP} + \mathcal{O}(M^{-2}).$$

In other words, and provided that M is sufficiently large, the values of the sequence are equispaced in $[0, P]$, up to $\mathcal{O}(M^{-1})$ (the "degradation" in the power of M is obvious, since we wish for a *uniform* bound and k itself is $\mathcal{O}(M)$). This proves equidistribution.   □

Equispacing is, of course, a considerably stronger feature than equidistribution. Thus, the angles χ_n are not just equidistributed in $[0, 2\pi]$, but the integers can be decomposed into a direct sum of sets $\{\mathcal{J}_M\}_{M=0}^{\infty}$ of consecutive integers such that each $\mathcal{J}_M$ is of length $\approx 2M$ and such that the sequence is asymptotically equispaced in each set. In other words, for large n, the sequence repeatedly traverses $[0, 2\pi]$ in sweeps of nearly-equispaced points, and the length of such sweeps increases linearily.

The aforementioned analysis applies only to $a = 0$. Let $\mathcal{I}$ be a measurable subset of $[0, 2\pi]$ of measure $m(\mathcal{I})$. The definition of equidistribution means that

$$\lim_{n \to \infty} \Pr\left(g_n \in \mathcal{I}\right) = \frac{m(\mathcal{I})}{2\pi},$$

consequently,

$$\lim_{n \to \infty} \Pr\left(g_n - an \in \mathcal{I}\right) = \lim_{n \to \infty} \Pr\left(g_n \in \mathcal{I} + an\right)$$
$$= \frac{m(\mathcal{I} + an)}{2\pi} = \frac{m(\mathcal{I})}{2\pi}.$$

Since this is true for every measurable subset, it follows that $\{g_n - an\}$ is also equidistibuted. Observe that, contrary to the situation in Section 4, the behavior does not hinge on the rationality – or otherwise – of a/π.

Having proved equidistribution (and noting the ubiquity of this behavior elsewhere in this article), it is only fair to conjecture that this persists for larger dimensions d. This is false!

Let $\mathbf{v}$ be an eigenvector of W with the eigenvalue λ. Unitarity implies that

$$\mathbf{v} = W^*W\mathbf{v} = \lambda W^*\mathbf{v},$$

consequently $\mathbf{v}$ is also an eigenvector of W^* with the eigenvalue λ^{-1}. Moreover,

$$\mathbf{0} = (W + W^*)\mathbf{v} = \left(\lambda + \frac{1}{\lambda}\right)\mathbf{v},$$

and (B5.4) implies that $\lambda^2 + 1 = 0$. Thus, $\sigma(W)$, the spectrum of W, consists of just $+i$ and $-i$, repeated as necessary. Moreover, if we insist on real W, clearly d must be even and the number of $+i$ and $-i$ must match. Simple calculation affirms that

$$W = \begin{bmatrix} 0 & -\alpha & -\beta & -\gamma \\ \alpha & 0 & \mp\gamma & \pm\beta \\ \beta & \pm\gamma & 0 & \mp\alpha \\ \gamma & \mp\beta & \pm\alpha & 0 \end{bmatrix}, \quad \alpha^2 + \beta^2 + \gamma^2 = 1,$$

is the most general form for real W, $d = 4$, consistent with (B5.4). Moreover, necessarily

$$W = Q^* \begin{bmatrix} -iI & O \\ O & iI \end{bmatrix} Q, \tag{B5.11}$$

where I is the $(d/2) \times (d/2)$ identity and Q is itself unitary (note, however, that whilst every unitary matrix Q in (B5.11) is consistent with (B5.4), not every choice yields a real W). We go back to the original map (B1.1) (with the special values of A and B):

$$\mathbf{z}_{n+1} = \left(A + \frac{1}{\|\mathbf{z}_n\|}B\right)\mathbf{z}_n = \left(I + \frac{1}{\|\mathbf{z}_n\|}W^*\right)A\mathbf{z}_n, \quad n = 0, 1, \ldots.$$

Letting $\tilde{\mathbf{z}}_n := Q\mathbf{z}_n$ yields, by virtue of unitarity of Q,

$$\tilde{\mathbf{z}}_{n+1} = \left(I + \frac{1}{\|\tilde{\mathbf{z}}_n\|}QW^*Q\right)QAQ^*\tilde{\mathbf{z}}_n$$

$$= \begin{bmatrix} (1 + i/\|\tilde{\mathbf{z}}_n\|)I & O \\ O & (1 - i/\|\tilde{\mathbf{z}}_n\|)I \end{bmatrix} QAQ^*\tilde{\mathbf{z}}_n.$$

Let $\tilde{A} := QAQ^*$ and note that it is unitary. Moreover, similarly to an earlier normalization, let

$$\tilde{\mathbf{u}}_n := \frac{\tilde{\mathbf{z}}_n}{\|\tilde{\mathbf{z}}_n\|} = Q\mathbf{u}_n, \quad n = 0, 1, \ldots.$$

As before, we stipulate that $\|\tilde{\mathbf{u}}_0\| = 1$ to obtain

$$\tilde{\mathbf{u}}_{n+1} = \frac{1}{\sqrt{n+2}} \left[\begin{matrix} \left(\sqrt{n+1}+i\right)I & O \\ O & \left(\sqrt{n+1}-i\right)I \end{matrix} \right] \tilde{A}\tilde{\mathbf{u}}_n,$$

an expression that involves just a single matrix A.

We denote by

$$E_\phi := \left[\begin{matrix} \cos\phi & \sin\phi \\ -\sin\phi & \cos\phi \end{matrix} \right]$$

the *Euler rotation* of the plane by the angle ϕ and set

$$\tilde{A} := \left[\begin{matrix} E_\alpha & O \\ O & E_\beta \end{matrix} \right].$$

It follows by induction that

$$\tilde{\mathbf{u}}_n = \left[\begin{matrix} e^{i\chi_n} E_{n\alpha} & O \\ O & e^{-\chi_n} E_{n\beta} \end{matrix} \right] \tilde{\mathbf{u}}_0.$$

In particular, we can see from our two-dimensional analysis that the iterants $\{\tilde{\mathbf{u}}_n\}$ (thus, also $\{\mathbf{z}_n\}$) are equidistributed on a direct product of two circles. This is a strict (in fact, lower-dimensional) subset of the surface of the sphere in $\mathcal{R}^4$, and equidistribution is confined to an infinitely small portion of the possible range of values.

References

[Halmos 1956] P. R. Halmos, *Lectures on Ergodic Theory*. Chelsea Publishing Co., New York, 1956.

[Körner 1988] T. W. Körner, *Fourier Analysis*. Cambridge University Press, Cambridge, 1988.

[Pólya and Szegő 1979] G. Pólya and G. Szegő, *Problems and Theorems in Analysis*. Springer-Verlag, Berlin, 1979.

[Rainville 1967] E. D. Rainville, *Special Functions*. Macmillan, New York, 1967.

[van Strien 1988] S. van Strien, "Smooth dynamics on the interval,"
    pp. 61–119, in: *New Directions in Dynamical Systems* (Bedford,
    T. and Swift, J., eds), London Math. Soc. Lecture Notes 127.
    Cambridge University Press, Cambridge, 1988.
[Varga 1962] R. S. Varga, *Matrix Iterative Analysis*. Prentice-Hall,
    Englewood Cliffs, N.J., 1962.

Part III
Historical Supplements

Foreword to
Historical Supplements

The short historical supplements that follow range over two and a half millennia and deal with certain aspects of spirals. They have been included to provide some feeling for the "texture" of the mathematics of different periods. For the complete discussions by these authors, the original documents should be consulted.

Having assembled these selections, and attempted some translations on my own, I doff my hat in acknowlegement of the scholars who have devoted years to the translation of and making sense of ancient mathematical documents. The manner of mathematical exposition changes rapidly; I have experienced the discomforture of changing goals, notations and modes of mathematical expression within my own professional lifetime. If ancient material has to be "watered up," I find that contemporary material often has to be "watered down" for my understanding.

I have placed a few comments of my own within the historical texts. These are designated by square brackets : [ ].

Historical Supplement A

From Plato: *Timaeus*

Excerpt from Plato's [427?–347 B.C.] *Timaeus*:

Now when all the stars which were necessary for the creation of time had attained a motion suitable to them, and had become living creatures having bodies fastened by vital chains, and learnt their appointed task, moving in the motion of the diverse, which is diagonal and passes through and is governed by the motion of the same, they revolved, some in a larger and some in a lesser orbit, and those which had the lesser orbit revolving faster, and those which had the larger more slowly. Now by reason of the motion of the same, those which revolved fastest appeared to be overtaken by those which moved slower although they really overtook them; for the motion of the same made them all turn in a spiral, and, because some went one way and some another, that which receded most slowly from the sphere of the same, which was the swiftest, appeared to follow it most nearly.

—*Timaeus*, 39. Translation by Benjamin Jowett.
Macmillan, London, 1892.

Historical Supplement B

From Plato: *Theaetetus*

Excerpt from Plato's *Theaetetus* (SOC. = Socrates. THEA. = Theaetetus):

> SOC. Let me offer an illustration [about the nature of knowledge in the abstract]: suppose that a person were to ask about some very trivial and obvious thing – for example, What is clay? and we were to reply, that there is a clay of potters, there is a clay of oven makers, there is a clay of brick makers; would not the answer be ridiculous?
>
> THEA. Truly.
>
> SOC. In the first place, there would be an absurdity in assuming that he who asked the question would understand from our answer the nature of 'clay', merely because we added 'of the image makers', or of any other workers. How can a man understand the name of anything when he does not know the nature of it?
>
> THEA. He cannot.
>
> SOC. Then he who does not know what science or knowledge is, has no knowledge of the art or science of making shoes?
>
> THEA. None.

SOC. Nor of any other science?

THEA. No.

SOC. And whan a man is asked what science or knowledge is, to give in answer the name of some art or science is ridiculous; for the question is 'What is knowledge?' and he replies, 'A knowledge of this or that.'

THEA. True.

SOC. Moreover, he might answer shortly and simply, but he makes an enormous circuit. For example, he might have said simply, that clay is moistened earth – what sort of clay is not to the point.

THEA. Yes, Socrates. there is no difficulty as you put the question. You mean, if I am not mistaken, something like what occurred to me and my friend here, your namesake Socrates, in a recent discussion.

SOC. What was that, Theaetetus?

THEA. Theodorus was writing out for us something about roots, such as roots of three or five, showing they are incommensurable by the unit: he selected other examples up to seventeen – there he stopped. Now as there are innumerable roots, the notion occurred to us to include them all under one notion or class.

SOC. And did you find such a class?

THEA. I think that we did; but I should like to have your opinion.

SOC. Let me hear.

THEA. We divided all numbers into two equal classes, those which are made up of equal factors multiplying into one another, which we compared to square figures and called square or equilateral numbers; that was one class.

SOC. Very good.

THEA. The intermediate numbers, such as three and five, and every other number which is made up of unequal factors, either of a greater multiplied by a less, or of a less multiplied by a greater, and when regarded as a figure is made up of unequal sides; all these we compared to oblong figures and called them oblong numbers.

SOC. Capital; and what followed?

THEA. The lines, or sides, which have for their squares the equilateral plane numbers, were called by us lengths or magnitudes; and the lines which are the roots of (or whose squares are equal to) the oblong numbers, were called powers or roots; the reason of this latter name being, that they are commensurable with the former [i.e., with the so-called lengths or magnitudes] not in linear measurement, but in the value of the superficial content of their squares; and the same about solids.

SOC. Excellent, my boys; I think that you fully justify the praises of Theodorus, and that he will not be found guilty of false witness.

THEA. But I am unable, Socrates, to give you a similar answer about knowledge, which is what you appear to want; and therefore Theodorus is a deceiver after all.

—*Theaetetus*, 147. Translation by Benjamin Jowett.
Macmillan, London, 1892.

Historical Supplement C

From Archimedes: *Peri Elikon*

Excerpts from Archimedes' [287–212 B.C.] *Peri Elikon* [On Spirals]. The brackets { } designate Heath's remarks. Heath, of course, has introduced modern notation.

Definitions.

1. If a straight line drawn in a plane revolve at a uniform rate about one extremity which remains fixed and return to the position from which it started, and if at the same time as the line revolves, a point move at a uniform rate along the straight line beginning from the extremity which remains fixed, the point will describe a spiral {'elix} in the plane.

2. Let the extremity of the straight line which remains fixed while the straight line revolves be called the origin {'archa: literally, the beginning of the spiral}.

. . .

Proposition 12.

If any number of straight lines drawn from the origin meet the spiral make equal angles with one another, the lines will be in arithmetical progression.

{The proof is obvious}

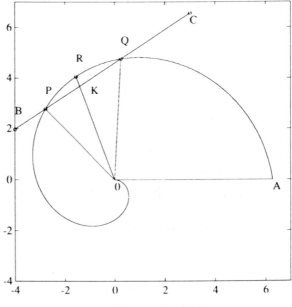

Figure 47.

Proposition 13.

If a straight line touch the spiral, it will touch it at one point only.

Let O be the origin of the spiral and BC a tangent to it. (See fig. 47.)

If possible, let BC touch the spiral in two points P, Q. Join OP, OQ, and bisect the angle POQ by the straight line OR meeting the spiral in R.

Then {Prop. 12} OR is an arithmetic mean between OP and OQ or

$$OP + OQ = 2OR.$$

But in any triangle POQ, if the bisector of the angle POQ meets PQ in K,

$$OP + OQ > 2OK.$$

{Known proposition. Assumed here.}

Therefore $OK < OR$, and it follows that some point on BC between P and Q lies within the spiral. Hence BC cuts the spiral, which is contrary to the hypothesis.

* * *

Proposition 24

The area bounded by the first turn of the spiral and the initial line is equal to one third of the 'first circle'.

$$\{= (\tfrac{1}{3}\pi)(2\pi a)^2, \text{where the spiral is } r = a\theta\}.$$

* * *

Let O be the origin, OA the initial line, A the extremity of the first turn.

Draw the 'first circle', i.e., the circle with O as centre and OA as radius.

Then, if C_1 be the area of the first circle, R_1 that of the first turn of the spiral bounded by OA, we have to prove that

$$R_1 = \tfrac{1}{3}C_1.$$

For, if not, R_1 must be either greater or less than C_1.

I. If possible, suppose $R_1 < \tfrac{1}{3}C_1$.

We can then circumscribe a figure about R_1 made up of similar sectors of circles such that, if F be the area of this figure,

$$F - R_1 < \tfrac{1}{3}C_1 - R_1,$$

whence $F < \tfrac{1}{3}C_1$.

Let $OP, OQ, \ldots$ be the radii of the circular sectors, beginning from the smallest. The radius of the largest is, of course, OA.

The radii then form an ascending arithmetical progression in which the common difference is equal to the least term OP. If n be the number of the sectors, we have {by Prop. 10, Cor. 1}

$$nOA^2 < 3(OP^2 + OQ^2 + \ldots + OA^2)$$

(see fig. 48) and, since the similar sectors are proportional to the squares on their radii, it follows that

$$C_1 < 3F,$$

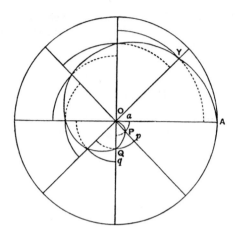

Figure 48.

or

$$F > \tfrac{1}{3}C_1.$$

But this is impossible, since F was less than $\tfrac{1}{3}C_1$.
Therefore

$$R_1 \not< \tfrac{1}{3}C_1.$$

II. If possible, suppose $R_1 > \tfrac{1}{3}C_1$.

We can then *inscribe* a figure made up of similar sectors of circles such that, if f be its area,

$$R_1 - f < R_1 - \tfrac{1}{3}C_1,$$

whence $f > \tfrac{1}{3}C_1$.

If there are $(n-1)$ sectors, their radii, as $OP, OQ, \ldots$, form an ascending arithmetical progression in which the least term is equal to the common difference, and the greatest term, as OY, is equal to $(n-1)OP$.

Thus {Prop. 10, Cor. 1},

$$nOA^2 > 3(OP^2 + OQ^2 + \cdots + OY^2),$$

whence $C_1 > 3f$, or $f < \tfrac{1}{3}C_1$; which is impossible, since $f > \tfrac{1}{3}C_1$. Therefore $R_1 \not> \tfrac{1}{3}C_1$.

Since then R_1 is neither greater nor less than $\frac{1}{3}C_1$,

$$R_1 = \tfrac{1}{3}C_1.$$

—Excerpted from the T. L. Heath edition of the
Works of Archimedes. See also Dijksterhuis,
pp. 275–277, and the supplemental
notes of W. R. Knorr, p. 435.

It is interesting to contemplate how a modern author would treat the proof of Proposition 13 into a statement about "spiral convexity."

Historical Supplement D

From Torricelli: *De Infinitis Spiralibus*

On the rectification of the logarithmic spiral. An excerpt from Evangelista Torricelli's *De Infinitis Spiralibus* [c. 1645]. This contains the first (known) construction of a line segment equal in length to the length of a curved arc. This precalculus result is based on the following geometrical identity.

Section 12.

THE LENGTH OF *BR* EQUALS THE LENGTH OF THE BROKEN LINE *BC ... L*. [See fig. 49.]

Starting from the longest ray AB of the spiral with center at A, and terminating with the shortest ray AL, we construct consecutively as many equal angles as we wish: $BAC, CAD,$ $DAF, \ldots, ZAL$ and complete the triangles $BAC, CAD,$ $DAF, \ldots, ZAL$.

Take segments AK and AV on lines AB and AC respectively equal to AL (i.e., to the shortest ray.) Thus the triangle AKV will be isosceles and hence its base angles $AVK = AKV$.

Let the prolongation of the line KV meet the prolongation of BC at the point R. I say that the length of BR is

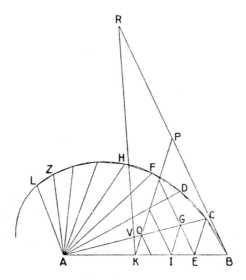

Figure 49.

equal to the sum of the lengths of all the inscribed segments
$BC, CD, DF, \ldots, ZL$.

In fact, because of the equality of the angles constructed
at the center $A, BA : AC = AC : AD$. [This follows from
Torricelli's definition of the logarithmic spiral given in the
first section of his manuscript.]

Thus the triangles BAC and CAD are similar, and the
same is true of all the remaining triangles up to the last
triangle which is ZAL. Now if we take alternatively on the
line AB and the line AC the segments $AE = AD, AG =
AF, AI = AH, AO = AZ$, all the way up until the last
which is AK, and which we have already taken equal to
AL, then according to the fourth proposition of Book I of
Euclid, the triangles AEC and ACD are congruent, as are
AEG and $ADF, \ldots$ We have

$$CE = CD, \ EG = DF, \ GI = FH, \ldots, ZL = OK.$$

Furthermore, the segments $BC, EG, \ldots$ will be parallel.
The remaining segments [taken alternately] CE, GI will also
be parallel.

For $BA : AC = AC : AE = AE : AG \ldots$

and therefore BC and EG are parallel lines. The same is true of the others.

We now prolong the last segment KO so that it intersects the line BR at the point P. Segment KP will be equal to the sum of all of the parallel segments $CE, GI, \ldots$:

$$KP = CE + GI + \cdots + OK.$$

For it is clear that if we prolong EG and all of the remaining segments which are parallel to it until they meet KP, these prolongations, will in fact divide the segment KP into as many parts as there are segments $CE, GI, \ldots, OK$ to which parts $CE, GI, \ldots, OK$ will be equal respectively.

Segment BP will also be equal to the sum of all of the parallel segments $BC, EG, \ldots$:

$$BP = BC + EG + \cdots.$$

This becomes clear if this time we prolong segments such as GI together with all the segments that are parallel to them until they intersect BP.

But it is now the case that $KP = PR$. This can be demonstrated as follows. Triangles KBR and VOK have equal angles (i.e., are similar) because ang KBR = ang ABC = ang ACD = ang ACD = ang ACE = $\ldots$ ang AOK = ang VOK. But BKR and OVK are both supplementary angles to the angles at the base of the isosceles triangle AKV. Thus angles $BKR = OVK$. Thus, for the remaining angles KRB and OVK must be equal, and therefore $KP = PR$.

Therefore $BR = KP + BP$. Thus BR will be equal to the sum of both the sums aforementioned. That is,

$$BR = BC + CE + \cdots,$$

and by what we proved above [i.e., $CE = CD, GI = FH, \ldots,$ $ZL = OK$] : BR must therefore be equal to the sum of the segments inscribed in the spiral:

$$BR = BC + CD + \cdots + ZL.$$

The excerpt just given was translated, rearranged and amplified from the Italian version of E. Carruccio.

Now notice what happens when the angle BAL is subdivided into more and more equal parts: The equal angles AKV and AVK approach a right angle, the point C approaches B, and the secant BC approaches the tangent to the spiral at B. Therefore we arrive at Torricelli's theorem:

Given a logarithmic spiral with center A. To obtain a line segment whose length equals that of the spiral arc from B to L: On the segment AB lay off AK equal in length to AL. At K erect a perpendicular to AB. At B construct the tangent to the spiral and allow this tangent to intersect the perpendicular at R. Then the length BR equals the length of the spiral arc BL.

Torricelli goes beyond. Allow the logarithmic spiral to wrap itself around infinitely often and approach its center A. At A erect a perpendicular to AB and at B construct the tangent to the spiral. Allow the tangent to intersect the perpendicular at T. The length of the segment BT on the tangent is the total length of the spiral as it wraps inward from B to A.

Historical Supplement E

From Bernoulli: *Opera Omnia*

Excerpt from a letter of Johann (Jean) Bernoulli [Bernoulli 1968], written in Basel on the 10th of January 1711, relating to the determination of the central force on a moving body in a resisting medium that is given by the product of the density and certain powers of the speed of the body.

. . .

PROBLEM

Find the central force required in order that a body move along a given curve in a medium whose density varies according to a given law and which resists the body proportionally to the product of the density and the speed raised to an arbitrary power.

[The solution given by Bernoulli follows.]

. . .

Remark II. Mr. Newton has made an oversight in his Proposition XVI on page 288 of his *Philosophiae Naturalis Principiis Mathematicis*, in which he says that if a body is attracted to one point by a central force whose magnitude is proportional to the reciprocal of the $n + 1$ st power of its distance from that point, and is moving in a medium whose density is the reciprocal of the nth power of that distance, that body will describe a logarithmic spiral whose pole or center will be the point towards which all the forces are directed. I have found, by my own analysis, that this is true only in the case when $n = 1$. This is the case of Prop. XV just prior to XVI.

For let x be the ray of the spiral, and let h be the secant of the constant angle that is made with each of its rays, ... Let c be the number whose logarithm is 1; f, the central force directed toward the center of the spiral; v' the speed of the body; R, the resistance of the medium and D its density. If one assumes, with Mr. Newton, that

$$D = 1 : x^n$$

and $R = mvvxD$, then according to my analysis one finds that we must have

$$f = x^{-3} \times c^{\pm 2mhx^{1-n} : (1-n)}$$

and not $f = 1 : x^{n+1}$, as Mr. Newton said, to have the body move in a logarithmic spiral. One concludes from this, that in contradistinction to the assertion $f = 1 : x^{n+1}$, the force will not equal a power of x except in the case when $n = 1$.

—Johannis Bernoulli, *Opera Omnia*, I, pp. 502-508.
Translated from the French with some liberties taken.[83]

Historical Supplement F

From Sylvester: *Note on the Successive Involutes of a Circle*

Excerpts from an 1868 article of James Joseph Sylvester.

... I shall use ϕ, s, r, θ to denote the angle of contingence, arc, radius, and vectorial angle of the curves under consideration.

. . .

Let now $\theta = 2\Phi, (a/2)r = \rho^2 \ldots$

then

$$\Phi = \sin^{-1}(1/2)a/\rho + (1/a)\sqrt{\rho^2 - (a/2)^2}$$

is the polar equation to a known curve (of the kind used by Captain Moncrieff in his barbette-gun carriage). It is of the class of curves generated by a fixed point on a wheel rolling on a plane. Such a curve may be termed the *convolute* of a circle of a *pitch* denoted by the ratio of the distance of the fixed point *below* the centre to the radius of the revolving circle; thus a convolute of zero pitch is the spiral of Archimedes, a convolute of unit pitch the first involute to the circle: the general equation to a convolute, when the distance below the centre is d and the radius a, is given by the Rev. James White

in the last September Number of the *Educational Times* and
is easily shown to be

$$\Phi = \sin^{-1}(d/\rho) + (1/a)\sqrt{\rho^2 - (d/2)^2}.$$

Similarly, we may define the pitch of the second involute to
be the ratio of the distance of its apse from the centre to the
radius; and then we are conducted to the observation that
whilst the convolute of the first pitch is the first involute,
the convolute of half pitch, on applying to it one of simplest
forms of M. Chasles's or Mr. Roberts method of transformation (given in Dr. Salmon's *Higher Plane Curves*, p. 236),
namely, doubling the vectorial angle and squaring the radius
vector, becomes converted into the second involute of half
pitch. Since for this curve

$$r = (a/2)(\phi^2 + 1) = ds/d\phi,$$

we see that it may be completely defined, without reference to
any theory of involutes, as the curve whose radius of curvature
at any point is equal to its radius vector reckoned from a given
origin. It is the curve which completely satisfies the equation
$rd\cos^{-1}(dr/ds) = s$ [*sic*], the two arbitrary parameters which
the complete integral of this equation should contain being
furnished by the linear magnitude and angle of swing of the
curve round the given origin.*

   *[Sylvester's footnote]: This evolute possesses the prop-
erty, which serves to characterize it completely, of cutting the
originating *circle* (its second evolute) orthogonally. For when
$r^2 = a^2, G^2 = 0$, that is, the tangent to the curve passes
through the centre. Moreover, since $G = 0$ gives $\phi = 1$, it follows that the curve cuts out of the circle an arc equal in length
to the diameter. Summarizing such of its principal properties
as have fallen in our way, we see that it bisects the line joining
the centre of the originating circle and the cusp of the first
involute; that it cuts the said circle orthogonally; that its radius of curvature is everywhere equal to its elongation from
the centre; that it is a trajectory to a central force varying as
the inverse cube of the shortest distance from the periphery
of the originating circle; that its arco-radial equation is only
of half the number of dimensions of the general involute of
the same order; and that by the simplest form of quadratic
transformation (namely, that which leaves unaltered the in-
clination of the tangent to the radius vector) it changes into
the half-pitch circular convolute; not to add that its polar

equation is even simpler than that of the first involute. Certainly, then, as it seems to me, it ought to take permanent rank among the spirals which have a specific name on the geometrical register; and for want of a better, with reference to the place where its properties first came into relief, it might be termed the *Norwich spiral*.

—From: "Note on the Successive Involutes to a Circle," J. J. Sylvester, *Mathematical Papers*, vol. II, pp. 638–9.

Historical Supplement G

From Poincaré: *Mémoire sur les Courbes*

This supplement excerpted from an 1882 article by Henri Poincaré relates to the nature of the singular points and to the limit cycles of the direction field of the solutions of first-order autonomous differential equations in the plane.

Chapter VI
The theory of limit cycles.

From what we have seen previously, the characteristics [orbits, trajectories] can be put into four categories:

1. Cycles [closed orbits; periodic solutions];

2. spirals that one can follow indefinitely in two directions without coming to a node or without turning around a spiral point [Fr.: foyer = focus = spiral point] and without coming back to the point of departure;

3. The characteristics that one can follow indefinitely in one direction without meeting a node or approaches a spiral point, but which, in the other direction, leads to a node or approaches infinitely closely to a spiral point;

4. Those that lead on both sides to a node or to a spiral point.

. . .

Chapter VII

Complete discussion of some examples.

Third example. Consider the equation

$$\frac{dx}{x(x^2 + y^2 - 1) - y(x^2 + y^2 + 1)}$$
$$= \frac{dy}{y(x^2 + y^2 - 1) + x(x^2 + y^2 + 1)}.$$

There is only one singular point [critical point] in each hemisphere; this is the point $x = y = 0$ and it is a spiral point.

There are no singular points on the equator which is a characteristic and is therefore a limit cycle. Consider the topographical system of circles whose centers are at the origin, that is to say, of the circles

$$x^2 + y^2 = \text{const.}$$

The contact curve of this topographical system is

$$(x^2 + y^2)(x^2 + y^2 - 1),$$

that is to say, all these circles are cycles without contact [non-tangential; traversal] except for the circle of radius 1 which is a limit cycle. [Here Poincaré seems to mean the following: if one rewrites the differential equation above in parametric form, one has $x' = DL, y' = DR$, where DL is the denominator on the left and DR is the denominator on the right, and where $'$ designates differentiation with respect to the parameter t. A slight algebraic manipulation of these two equations yields

$$(x^2 + y^2)' = 2(x^2 + y^2)(x^2 + y^2 - 1),$$

from which his conclusion follows.] There are no other limit cycles. The system of characteristics therefore has the aspect of the following figure.

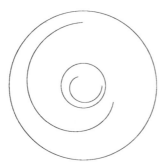

Figure 50.

Fourth example: Consider the equation

$$\frac{dx}{x(x^2 + y^2 - 1)(x^2 + y^2 - 9) - y(x^2 + y^2 - 2x - 8)}$$

$$= \frac{dy}{y(x^2 + y^2 - 1)(x^2 + y^2 - 9) + x(x^2 + y^2 - 2x - 8)}$$

We see that there are three singular points:
1. The point $O : x = y = 0$;
2. The points A and B at the intersection of the circles

$$x^2 + y^2 - 9 = 0, x^2 + y^2 - 2x - 8 = 0.$$

The point O is a spiral point; the point A is a node; the point B is a saddle point. We see that as in the preceding example, the equator is a limit cycle; that the circles whose centers are at the origin are cycles without contact, except for the circles $x^2 + y^2 - 1 = 0, x^2 + y^2 - 9 = 0$, and these are characteristics.

The first of these, which doesn't go through a singular point, is a limit cycle; the second goes through a node and a saddle point.

There are therefore three kinds of characteristic: the first kind twist around the spiral point O and have

$$x^2 + y^2 - 1 = 0$$

for a limit cycle; the second lead to the node A and have

$$x^2 + y^2 - 1 = 0$$

as a limit cycle; the third lead to the node A and have the equator as a limit cycle.

The following are exceptional characteristics:

1. The equator;
2. The circle $x^2 + y^2 = 1$;
3. The circle $x^2 + y^2 = 9$;
4. A characteristic that starts from the saddle point B and has the equator as a limit cycle.

5. A characteristic that starts from the saddle point B and has $x^2 + y^2 = 1$ as a limit cycle.

— From: Henri Poincaré, "Mémoire sur les courbes définies par une équation différentielle," *Journal de Mathématiques Pures et Appliquées*, v. 8 (1882), Chap. VII. There are three other articles in this memoir: v. 7 (1881), 375–422; v. 1 (1885), 167–244; v. 2 (1886), 151–217.

Historical Supplement H

From Hlawka: *Gleichverteilung und Quadratwurzelschnecke*

The excerpts that follow constituted the original inspiration for these Hedrick lectures. They are taken from Edmund Hlawka's 1980 article on the equidistribution (also referred to as "uniform distribution") of the angles of the square root spiral.

Let us consider the nth triangle $OP_{n-1}P_n(n = 1, 2, \ldots)$ in the square root spiral, i.e., [spiral of Theodorus]. See Figure 51.

Since the angle at P_{n-1} is a right angle,

$$\sin \alpha_n = 1/\sqrt{n+1}.$$

Therefore, φ_n, the angle between the ray OP_0 and OP_n, is given by $\varphi_n = \alpha_1 + \cdots + \alpha_n$. Therefore we have

$$\varphi_n = \sum_{k=2}^{n+1} \arcsin(1/\sqrt{k}).$$

W. Neiss [1966, pp. 241–43] considered the sequence $\bar{w} = (\varphi_n)$ and proved that the sequence $w = (\frac{1}{2}\pi)\bar{w}$ is equidistributed modulo 1.

$\cdots$

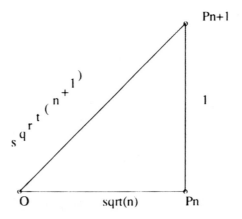

Figure 51.

A sequence of real numbers β_n is called equidistributed if the following is the case: Consider all subintervals I of the half-open unit interval $0 \leq \xi < 1$. Next, consider the first N terms of the sequence $\beta_1 - [\beta_1], \ldots, \beta_N - [\beta_N]$ (where $[\beta]$ is the Gauss bracket; i.e., the largest integer $\leq \beta$) and count how many of these terms lie in I. Designate by $A(N, I)$ the number of such terms, and by $A^*(N, I) = A/N$, the relative frequency. If, for every I, the limit as $N \to \infty$ of $A^*(N, I)$ exists and is equal to $\lambda(I)$, the length of the interval I, then the sequence β_n will be called equidistributed modulo 1. From the theorem of Neiss that the sequence $(1/2\pi)(\varphi_n)$ is equidistributed, we may conclude the following: Let $J =< \rho, \sigma <$ be a subinterval of $< 0, 2\pi <$ (more generally, J can be an arbitrary interval modulo 2π, that is, it can split into two subintervals of $< 0, 2\pi <$; then with J we associate the subinterval $I = (\frac{1}{2\pi})J$ of the unit interval. Consider the first N rays and let $N'(J)$ be the number of that lie in the sector bounded by the rays ρ and σ. Then we have $N'(J) = A(N, I)$. Therefore, from the equidistribution of the sequence $(\frac{1}{2\pi})\varphi_n$, we conclude that

$$\lim_{N \to \infty} \frac{N'(J)}{N} = \frac{1}{2\pi}\lambda(J).$$

We shall say that the sequence of rays φ_n is equidistributed modulo 2π.

Neiss' proof is elementary but not simple. We shall give another proof (which the author already gave in his 1968 lectures on selected chapters of the theory of numbers.) In doing so we will have accomplished rather more in that we will have demonstrated how modern methods of analytic number theory and of the theory of uniform distributions lead us to deeper things. It would also be easy to consider sequences (φ_n) that are more general than the specific sequence defined by (1) below.

Since the author of this work is (as is Theodorus) one of the older mathematicians, we will explain the method with the specific sequence ω. Our point of departure is the asymptotic expansion

$$\varphi_n = 2\sqrt{n+1} + K + \frac{1}{6\sqrt{n+1}} + O(n^{-3/2}), \qquad (1)$$

where K may be called the "square root spiral" constant. Its value is $\leq 0,75$.

We shall deepen the theorem of Neiss by obtaining bounds for the discrepancy $D_N(\omega)$. By this symbol we mean

$$D_N(\omega) = \sup |A^*(N, I) - \lambda(I)|$$

[where the sup is taken over I].

We shall prove in Theorem 1 that

$$D_N(\omega) \leq \frac{120}{\sqrt{N}}. \qquad (2)$$

In the other direction, we shall prove that there is a constant c_1, which could be calculated, such that

$$D_N(\omega) \geq \frac{c_1}{\sqrt{N}}. \qquad (3)$$

In addition to the sequence (φ_n), we shall also consider the sequence $\omega(L) = (\varphi_{n+L} - \varphi_L)$, where $L \geq 1$, and we shall prove that for this sequence,

$$N \cdot D_N(\omega(L)) \leq C_2 \sqrt{N + L}$$

$$N \cdot D_N(\omega(L)) \geq C_3 \sqrt{N + L}.$$

The sequence ω is therefore not "uniformly equidistributed."

In Theorem 4 [not reproduced here] we shall prove the

following. Let $p_1, p_2, \ldots, p_s$ be distinct prime numbers. Then the s-dimensional sequence $\frac{1}{2\pi}(\varphi_{np_1}, \ldots, \varphi_{np_s})$ is equidistributed modulo 1 in R^s. This is to be understood as follows. Let I be an s-dimensional subinterval of the unit hypercube. Consider the first N points

$$\left(\beta_{np_1} - [\beta_{np_1}], \ldots, \beta_{np_s} - [\beta_{np_s}]\right), \left(\beta_n = \frac{1}{2\pi}\varphi_n\right)$$

and designate by $A(N, I)$ the number of these points that lie in I. Then we shall have

$$\lim_{N \to \infty} \frac{1}{N} A(N, I) = \lambda(I), \tag{4}$$

where $(\lambda(I)) =$ the volume of I.[84]

As a rough upper bound on the discrepancy

$$D_N(\omega) = \sup_I \left| \frac{1}{N} A(N, I) - \lambda(I) \right|$$

we obtain

$$D_N \leq c_5 N^{-2^{-s}} (\log N)^{s-1}.$$

But if we use a deep theorem of W. Schmidt [1970, pp. 189–201], we can obtain

$$D_N \leq c_6 N^{-[2(1+\varepsilon)(s+1)]^{-1}}.$$

At the end, we will turn to another type of question. In a nice piece of work, E. Teuffel [1959] posed and solved the following problem, which was raised independently by Neiss:

Is it possible to solve the equation

$$\varphi_{n+L} - \varphi_n = g$$

in positive integers g, L, n? By using a theorem of Sylvester–Schur [given two positive integers n and s with $s \leq n$, then there is a prime $p > s$ and an integer a with $n \leq a \leq n + s$, which is divisible by p. I. Schur: *Gesammelte Werke*, vol. 3, p. 140], we have succeeded in simplifying Teuffel's proof of his theorem that this is not possible. [This implies that no other P_n's can lie on the line joining O and P_n.] We also succeeded in dealing in a simple manner with the problem, raised by Teuffel, whether the equation

$$\varphi_{n+L} - \varphi_n = \arctan\sqrt{\frac{L}{n+1}}$$

is solvable in positive integers. [This implies that no other P_n's can lie on the line joining P_n and P_{n+1}.]

$$\cdots$$

§1. In order to arrive at (1), we use the Euler summation formula in the following form:

$$\sum_{k=2}^{n+1} g(k) = \frac{1}{2}(g(2)+g(n+1)) + \int_2^{n+1} g(x)dx + K_0 + R_{n+1}, \quad (5)$$

where

$$K_0 = \int_2^\infty \varrho(x)g''(x)dx, \quad (6)$$

$$R_{n+1} = \int_{n+1}^\infty \varrho(x)g''(x)dx, \quad (7)$$

and where $g(x) = \arcsin\frac{1}{\sqrt{x}}, \varrho(x) = \frac{1}{2}\{x\}(1-\{x\})$, and $\{x\} = x - [x]$ and are defined for all x in $2 \le x \le n+1$.

Since we have

$$g'(x) = \frac{1}{2}(x\sqrt{x-1})^{-1}, \quad g''(x) = \frac{1}{2}(3x-2)(x^2(x-1)^{3/2})^{-1},$$

then we have

$$|R_{n+1}| \le \int_{n+1}^\infty g''(x)dx = g'(n+1),$$

so that

$$R_{n+1} = \frac{1}{2}\theta n^{-3/2}. \quad (8)$$

(Here and in what follows, θ will always designate numbers of absolute value ≤ 1.) From this bound, one can also conclude the existence of the number K_0. Moreover, we have

$$G(x) = \int g(x)dx = x\arcsin\frac{1}{\sqrt{x}} + \sqrt{x-1}.$$

To compute $G(n+1)$, we can insert $x = (n+1)^{-1/2}$ in the expansion

$$\arcsin x = x + \tfrac{1}{6}x^3 + \left(\tfrac{5}{2}\right)^4 \theta x^5, \tag{9}$$

leading to

$$(n+1)\arcsin \frac{1}{\sqrt{n+1}}$$
$$= \sqrt{n+1} + \frac{1}{6\sqrt{n+1}} + \left(\frac{5}{2}\right)^4 \theta (n+1)^{-3/2}. \tag{10}$$

Moreover, inserting the same value of x in the expansion

$$\sqrt{1+x} = 1 + \frac{x}{2} + \frac{x^2}{4\sqrt{1-x}}\theta, \tag{11}$$

one obtains

$$\sqrt{n} = \sqrt{n+1} - \frac{1}{2\sqrt{n+1}} + \frac{\theta}{8(n+1)^{3/2}}. \tag{12}$$

We therefore get

$$\varphi_n = 2\sqrt{n+1} + K + \frac{1}{6}\frac{1}{\sqrt{n+1}} + \frac{1}{4}5^4 n^{-3/2}, \tag{13}$$

where

$$K = \frac{1}{2}\arcsin\frac{1}{\sqrt{2}} + K_0. \tag{14}$$

Therefore $K = \frac{\pi}{8} + K_0, s$, where $K_0 \leq \frac{1}{2}g'(2) = \frac{1}{4}$. It follows that we have $K \leq 0,75$. Thus,

$$r_n = \sqrt{n+1} = \frac{1}{2}\varphi_n - K + O\left(\frac{1}{\sqrt{n}}\right).$$

Thus, to a first approximation, we are dealing with a spiral of Archimedes, a fact already noticed by Teuffel [1958]. We now compare the sequence ω with the sequence $\tilde{\omega} = (\psi_n)$, where

$$\psi_n = 2\sqrt{n+1} + \frac{1}{6\sqrt{n+1}} + K. \tag{15}$$

From (11) we have

$$\varepsilon_n = |\varphi_n - \psi_n| < 3 \cdot 10^2 n^{-3/2}. \tag{16}$$

Now it is well known that for all values of $c > 0$ the sequence $(c\sqrt{n})$ is equidistributed modulo 1. So in particular, the sequence $\omega_1 = (2\sqrt{n})$ [when reduced mod 1] is equidistributed. Since the sequence $\tilde{w}$ differs from ω_1 only by a convergent sequence, it follows that $\tilde{w}$ is equidistributed. Moreover, ω and $\tilde{\omega}$ differ by a null sequence. Therefore it follows that ω is also equidistributed. This proves the theorem of W. Neiss. For this purpose, we hardly needed so exact a representation as that provided by (13). The representation

$$\varphi_n = 2\sqrt{n+1} + K + \frac{\theta}{\sqrt{n+1}}$$

would have sufficed. However we need (13) to determine the exact order of the discrepancy $D_N(\omega)$. The discrepancy $D_N(\omega)$ is defined by

$$\sup_J \left| \frac{N'(J)}{N} - \frac{\lambda(J)}{2\pi} \right|, \tag{17}$$

where J is an arbitrary subinterval $< \varrho, \sigma <$ of $< 0, 2\pi <$ and where $N'(J)$ is the number of the $\varphi_n (1 \leq n \leq N)$ that lie in J. The quantity $D_N(\tilde{\omega})$ is defined analogously. We will next compare the two quantities $D_N(\omega)$ and $D_N(\tilde{\omega})$ and make use of the following theorem which was established by the author (See [Kuipers and Neiderreiter 1974, Chap. 2, Theorem 4.1, p. 32] but stated there mod 1.)

For every $\varepsilon > 0$, we have

$$\Delta_N = |D_N(\omega) - D_N(\tilde{\omega})| < \varepsilon + \frac{\overline{N}}{N}, \tag{18}$$

where $\overline{N}$ is the number of the ε_n with $1 \leq n \leq N$ and $\varepsilon_n > \varepsilon$. Now, from (16) we have $\overline{N} \leq \overline{\overline{N}}$, where $\overline{\overline{N}}$ is the number of n with $1 \leq n \leq N$ and $3 \cdot 10^2 n^{-3/2} > \varepsilon$. Therefore we have

$$\overline{N} \leq \overline{\overline{N}} \leq 40\varepsilon^{-2/3}. \tag{18'}$$

It therefore follows that

$$\Delta_N < \varepsilon + 40N\varepsilon^{-2/3}. \tag{19}$$

Now take $\varepsilon = (\frac{40}{N})^{3/5}$, then with

$$c_1 = 2 \cdot 40^{3/5} \tag{20}$$

we obtain

$$\Delta_N = |D_N(\omega) - D_N(\tilde{\omega})| < c_1 N^{-3/5}. \tag{21}$$

§2. We now have to determine $D_N(\tilde{\omega})$ through (18). To this end it suffices to consider intervals of the form $< K, \gamma + K <$ where $0 < \gamma < 2\pi$. Let I be such an interval, then we have to bound the number of solutions of

$$K \le \psi_n - 2\pi \left[\frac{\psi_n}{2\pi} \right] < \gamma + K \tag{22}$$

with $1 \le n \le N$. Call this number $N'(I)$. Set $\gamma = 2\pi\beta$,

$$g_n = \left[\frac{\psi_n}{2\pi} \right], \beta_n = \frac{1}{2\pi} \psi_n,$$

then from (15),

$$\beta_n = \sigma(n+1), \tag{23}$$

where

$$\sigma(x) = \left(\sqrt{x} + \frac{1}{12\sqrt{x}} + \frac{K}{2} \right) \frac{1}{\pi}. \tag{23'}$$

Now let $\overline{K}_1 = K/2\pi$. Then we have to study the inequality

$$\bar{K}_1 + g_n \le \beta_n < g_n + \beta + \bar{K}_1 \tag{24}$$

for $1 \le n \le N$. Since the derivative σ' is always positive, we have

$$g_n = [\beta_n] \le [\beta_N] = g_N = g. \tag{25}$$

We have $g_N \ge g_0 + 1$ where

$$g_0 = \left[\frac{2\sqrt{N+1} + K}{2\pi} \right] - 1 \tag{26}$$

Now also

$$\beta_N \ge \frac{2\sqrt{N+1} + K}{2\pi} \ge \left[\frac{2\sqrt{N+1} + K}{2\pi} \right] = g_0 + 1.$$

Therefore, if we set $g_N = g$,

$$N'(I) \le \sum_{L=1}^{g} a(L, I) = N_1'(I), \tag{27}$$

where $a(L, I)$ designates the number of solutions of

$$\bar{K}_1 + L \leq \beta_n < L + \beta + \bar{K}_1 \qquad (28)$$

with $1 \leq n \leq N$. Naturally the number a can also be equal to zero. On the other hand,

$$N'(I) \geq \sum_{L=0}^{90} a(L, I) = N_2'(I). \qquad (29)$$

We have arrived at (28), and therefore we must discuss the inequality

$$\bar{K}_1 + L \leq \frac{1}{\pi} \left(\sqrt{n+1} + \frac{1}{12\sqrt{n+1}} + \frac{K}{2} \right) < L + \beta + \bar{K}_1,$$

that is to say, the inequality

$$a_1 = \pi L \leq \sqrt{m} + \frac{1}{12\sqrt{m}} < (L + \beta)\pi \leq a_2, \qquad (30)$$

where we have written $n + 1 = m$. Now m satisfies the right-hand side of the inequality precisely if

$$\sqrt{m} \leq \tfrac{1}{2}(a_2 + \sqrt{a_2^2 - \tfrac{1}{3}})$$

then $L \geq 1$ and therefore $a_1 > 3$. One gets a lower bound for $\sqrt{m}$ if one replaces a_2 by a_1. Furthermore, for every $r \geq 2$, we have

$$\sqrt{1 - \frac{1}{3r^2}} = 1 - \frac{1}{6r^2} + \frac{\theta}{36r^4}\sqrt{\frac{12}{11}}, \qquad (*)$$

(cf. (11)). Therefore, if we apply $(*)$ with $r = a_1$ and $r = a_2$, we obtain

$$a_1 - \frac{1}{12a_1} - \frac{\theta}{12a_1^3} \leq \sqrt{m} < a_2 - \frac{1}{12a_2} + \frac{\theta}{12a_2^3},$$

from which, introducing abbreviations b_1 and b_2,

$$b_1 = a_1^2 - \frac{1}{6} - \frac{5}{a_1^2} \leq m < a_2^2 - \frac{1}{6} + \frac{5}{a_2^2} = b_2.$$

Therefore, $a(L, I) = [b_2] + [-b_1] + 1$, and consequently,

$$a(L, I) \leq 2\beta L\pi^2 + \beta^2 \pi^2 + \frac{10}{L^2\pi^2}\theta + 1. \qquad (31)$$

It follows that

$$N_1'(I) \leq \pi^2 \sum_{L=1}^{g} \left(2\beta L + \beta^2 + \frac{1}{L^2}\theta + 1\right),$$

so that

$$N_1'(I) \leq \pi^2 \beta g(g+1) + \beta^2 \pi^2 g + \frac{\pi^4}{6} + g. \qquad (32)$$

As a result, then, of (25),

$$g = g_N = \left[\frac{\sqrt{N+1}}{\pi} + \frac{K}{2\pi} + \frac{1}{12\pi\sqrt{N+1}}\right],$$

so that

$$N_1'(I) \leq \beta N + 20\sqrt{N} + 5.$$

On the other hand, $[b_2] + [-b_1] \geq b_2 - b_1 - 2$,

$$N_1'(I) \geq \pi^2 \sum_{L=1}^{g_0} \left(\left(2\beta L + \beta^2 - \frac{1}{L^2}\theta\right) - 2\right).$$

It follows from this that

$$N_1'(I) \geq \pi^2 \beta g_0(g_0+1) + \beta^2 \pi^2 g_0 - \frac{\pi^4}{6} - 2g_0\pi^2. \qquad (32')$$

Now we have

$$g_0 \geq \frac{\sqrt{N}}{\pi} + \frac{K}{2\pi} - 1,$$

so that

$$N_2'(I) \geq \beta N - 10\sqrt{N} - 5.$$

Using (27) and (29), we arrive at

$$|N'(I) - \beta N| \leq 20\sqrt{N} + 5.$$

Now since β was arbitrary, it follows that

$$D_N(\tilde{\omega}) \leq \frac{40}{\sqrt{N}}. \qquad (33)$$

Now, using (21), we get

$$D_N(\omega) \leq \frac{40}{\sqrt{N}} + \frac{80}{N^{3/5}},$$

and we finally arrive at:
Theorem 1:

$$D_N(\omega) \le \frac{120}{\sqrt{N}}.$$ (34)

...

—Excerpted from: E. Hlawka, *Gleichverteilung und Quadratwurzelschnecke*, Herrn Prof. R. M. Redheffer in Freundschaft zum 60. Geburtstag gewidmet. *Monatshefte für Mathematik*, **89** (1980), 19–44.

Epilogue

I

> You cannot find out what a man means by simply studying
> his spoken or written statements even though he has spoken
> or written with perfect command of the language and per-
> fectly truthful intention. In order to find out his meaning,
> you must also know what the question was (a question in his
> own mind and presumed by him to be in yours) to which the
> thing he has said or written was meant as an answer.
>
> —R. G. Collingwood, *An Autobiography*, p. 31.

R. G. Collingwood, an historian and philosopher, wrote these
words as a response to a recurrent event in his early professional life: he had to walk by the Albert Memorial every day on
his way to work, a monument that he considered to be "mis-
shapen, corrupt, crawling, verminous." Why, he asked himself,
had Scott, the architect, laid out a "thing so obviously, so incon-
trovertibly, so indefensibly bad?" Out of this questioning grew
a philosophy that was anti-logicist and which questioned a view
of knowledge, then popular in British philosophical circles, that
had grown out of the logicist position.

The spiral, a geometric concept of somewhat vague compass that often inspires wonder and delight, is found in nature, in art, in decoration, in myth, in legend, in illusion, in religious symbolism, and, of course, in mathematics both pure and applied. This small volume on spirals is an assemblage of theorematic material both old and new, theorematic conjecture, computer experience, computer graphics, historical monuments of mathematics that span two and one half millennia, historical and philosophical animadversions. Much of the mathematics it contains I believe is beautiful, stimulating, and often of practical interest. In the process of composing and editing this material, I have had to pass it before my eyes many times, and this review, particularly of the historical documents, has led me to marvel at the continuity that they imply and to appreciate how this continuity testifies to the continued presence of a living mathematical tradition, which sometimes has flickered low but has always managed to restore itself to full flame.

I began to wonder whether I could ever understand the materials of the past in the sense in which those materials were understood by the individual authors who composed them out of experiences and out of intellectual milieus and sensibilities that were far different than mine. I performed a thought experiment. I imagined that a well-eyed intellect from Quasar X-9 looked over this material, both the symbolic and the graphic. Would this intellect make of it what I make of it; would this intellect be able to bind together the visual and the verbal as I do? I believe the answer to these questions is "No." Let us not think that though we say we have interpreted these materials in a way that seems to us both accurate and adequate, we are reading the same article

> as the contemporaries of the authors. Perhaps the qualities they prized most highly are those which escape us; others which they hardly perceived may affect us deeply. ... The change of century, which means a change of reader, is comparable to an alteration in the text itself, always incalculable and always unforseen.

> —Paul Valéry, *Adonis*

Reuben Hersh has called attention to certain normative "myths" of mathematics. Among them are the myths of unity and universality that state:

> There is only one mathematics, indivisible, now and forever. ...Mathematics as we know it is the only mathematics there can be. If the little green men (and women?) from Quasar X-9 sent us their math textbooks, we would find again $A = \pi r^2$.

While admitting the utility of the various myths, Hersh questions that of universality by asking:

> What would it mean to talk about their (i.e., the little green beings) literature or art or mathematics? The very notion of comparing presupposes beings enough like us to make communication conceivable. But then, the possibility of comparison is not universal; it's conditional on their being 'enough like us'.

It is part of the myth structure of mathematics that total formalization is a possible; that the mathematical substance can be reduced to a sequence of symbols, of zeros and ones, if you like, and that such a reduction is not only possible but is a sufficient, self-standing statement of its own being. However, such a reduction would not be mathematics as it is created, intuited, discussed, studied, understood, interpreted or applied.

One might ask – one should ask: "What were the questions to which the historic monuments on display here were the answers?" When the questions are sharply focused, the answers may stand out clearly. Thus: the Prince died and Scott received a commission to do a monument. Result: the Albert Memorial. Though this simplistic answer would surely not have satisfied Collingwood, when we turn to the mathematical record, we often find that clear formulations even at this level are avoided. It is useful, then, to review briefly the historical selections placed in this book.

A. Plato's *Timaeus*:

> Those stars which revolved fastest appeared to be overtaken by those which moved slower although they really overtook

them; for the motion of the same made them all turn in a spiral, and, because some went one way and some another, that which receded most slowly from the sphere of the same, which was the swiftest, appeared to follow it most nearly.

How do the planets appear to move? An answer is part of descriptive astronomy. Even though the descriptions are not given a precision that might have been possible at the time, one detects the mathematical spirit at work here. The relative motions of the planets is made vivid, and the word spiral seems to be employed to describe combination of daily and annual rotation, of, say, the sun as it spirals up to the Summer Solstice and down to the Winter Solstice.

B. Plato's *Theaetetus*:

What is the nature of abstract knowledge? Plato's concern with square roots is in the service of creating a theory of abstraction. It has no prima facie relationship to spirals. This connection was made centuries later by Anderhub, though the figure he envisioned might very well have been drawn by the mathematicians of Plato's day.

The connection that I have made in these lectures and notes is this: given that abstraction is the life blood of the mathematical process, it seemed appropriate to discuss the problems involved in abstracting the notion of a spiral.

C. Archimedes' *Peri Elikon*:

Archimedes introduces his spiral into the mathematical literature and answers, among other things, the question of how much area is swept out by a ray that traces the spiral. Archimedes, undoubtedly the greatest mathematician and physicist of classical antiquity, discusses his spiral largely as an object of pure mathematics, using methods which in some measure anticipate those of integral calculus. He puts the spiral to the use of angle trisection and circle quadrature. Perhaps his interest in spirals was related to "the screw of Archimedes," a hydraulic device wherein a helical tube is used to raise water. The spiral is defined mechanically, and Archimedes in *The Method* has found it important to distinguish between a mechanical approach to matters and a purely mathematical one, assigning a higher status to the latter.

D. Torricelli's *De infinitis spiralibus*:

Find the length of the arc of the logarithmic spiral. In precalculus days, this was a difficult problem for general curves; and even for the circle, aspects of the problem were not cleared up until the 1880s. For the ellipse, the question led to a major chapter in the history of mathematical analysis: the theory of elliptic functions.

Comparing Torricelli's text with that of Archimedes, one arrives at the conclusion that this piece of work would have been quite within the capabilities of the Archimedean Age.

E. Jean Bernoulli's *Letter*:

We are now located squarely within the age of Newtonian celestial mechanics and the calculus. The question to be answered is put by the author clearly:

> Problem: Find the central force required in order that a body move along a given curve in a medium whose density varies according to a given law and which resists the body proportionally to the product of the density and the speed raised to an arbitrary power.

Spirals play a role as orbits under different central force laws.

F. James Sylvester's *Note on the Successive Involutes to a Circle*:

The very particular question leading to the spiral of Norwich is: find the curve whose radius of curvature at each point equals its distance to the origin. This is immersed in a large study of curves that roll on other curves and of the iteration of this rolling process. Lurking in the back of Sylvester's mind is Captain Moncrieff's application of these ideas to the design of gun carriages. The more general background is that of mid-Victorian kinematic technology with its mechanical linkages, gear trains, and so on. We may recall that a linkage problem in steam-engine design led the great Russian mathematician Pafnuty Lvovich Tschebyscheff to the notion of best approximation by polynomials and to his fundamental theorem in that field.

Hearkening back to the spirit of Bernoulli, Sylvester very eloquently (but I think with tongue in cheek) makes great claims for the spiral of Norwich. Where indeed, on average, are the great claims of yesteryear?

G. From Henri Poincaré's *Mémoire sur les courbes définies par une équation différentielle*:

Problem: Classify the behavior of a system of first order autonomous differential equations at its singular points. In two dimensions, the logarithmic spiral emerges as one form of prototypical behavior. Poincaré, who wrote profound treatises on celestial mechanics and its differential equations, takes us back to the movement of the heavens. The seeds of doubt as to the philosophic status of a deterministic universe will sprout in Poincaré's later work inasmuch as Newton's laws can lead to chaotic orbits.

It might be thought a denigration of the work of this Grand Master of Mathematics to have reprinted here only some examples that he gave in this great memoir, examples that might very well be used in an undergraduate course in differential equations. However, I recall the remark of the symphonic conductor George Solti to the effect that some of Mozart's work is too simple for children and too difficult for professionals.

H. From Edmund Hlawka's *Gleichverteilung und Quadratwurzelschnecke*:

Theodorus' spiral is now perceived to exhibit traces of mild chaos. The problem is: describe in some detail the chaotic distribution of the angles of the vertices of this spiral and in this way, extract some order out of the chaos.

The historical development of the theory of equidistributed sequences that stands behind the particular results of this paper has been described in great depth in the article by Hlawka and Binder. In the period from 1909 to 1916, many famous mathematicians, P. Bohl, H. Bohr, G. H. Hardy, W. Sierpinski, H. Weyl among them, contributed to this very beautiful union of analysis and number theory. The beginnings of this theory can be traced as far back as the 1780s when Joseph-Louis Lagrange raised questions about the existence of an "average motion" of planets whose orbits are described by nonharmonic trigonometric polynomials.

These few words give narrow indications of the questions that elicited the historical documents reprinted. In some cases, a wider scientific context of the question has been ventured. These wider contexts could be extended to a full scale historical study of the role of spirals. With the exception of the selections by Sylvester and Hlawka, all the material, if presented today, would be reformulated in a contemporary style and handled routinely by what would now be regarded as the elementary portions of college mathematics.

On the other hand, if Collingwood's "question–answer" demand is interpreted in a wider sense than above, we need not only know the question that was answered by the mathematical symbols, but, indeed, we need to know the whole contents of the particular mind and the particular experience from which it derived. Mathematical historiography has a long way yet to go toward that goal, and it is a goal that may never be reached. The tacit knowledge, the experiences and intuitions of the individual authors are not a part of the written record, and this material, the thoughts, conscious or unconscious, which are real enough, are considered by many not to constitute part of mathematics as such.

The agreed upon meaning and validity of the mathematical documents attested by the summaries above and often arrived at by a long and unrecorded process of social negotiation, display only the tip of the creative iceberg; it is part of what we understand by the communication of the mathematics that exists in the archival sense. The meaning and seeming continuity of several documents in historic sequence (illustrated in the above interpretations) are, in part, an artifact of our own contemporary desire and ability to create and enforce such continuity. We have managed to filter out all that does not suit our current vision. The ability to do so derives from much more than the naked archival mathematical residue.

In retrospect, if one were asked to point to a common underlying concern in articles reprinted above, one might, with some justification answer: the Cosmic Dance. The heavens and all therein twist and turn, now coalescing, now receding; now organizing themselves into spiral galaxies and now into patterns of scattered dust. The search for precise descriptions of the celestial choreography will go on for some time to come and will be answered by many more lines of mathematical symbols.

Allied to description is the possibility of prediction on the grand scale: Where are we going; and the possibility of postdiction: Where did we come from? Poincaré was one of the first post-Newtonians to raise his eyebrows at the strict determinism that Newtonian mechanics seemed to imply, and today, though answers to these questions are given regularly in the Sunday Supplements, the matter is moot.

II

In examining the material of this book, what also struck me forcibly was the difference between the formulaic material and the graphical material. Regarding these two kinds of material as pure blobs of ink spat out by a laser printer, how different in visual and semiotic quality they appear to be. Each type can stand alone, can stand in its own right and need not call upon the other for justification. Yet, by some miracle of thought, the mind can fuse the two and bind them together intimately into a certain totality. It is part of our heritage from the Greek mathematicians that simple questions about shape have led to deep symbolic results or to deep processes of computation. The French author Paul Valéry put it this way:

They (the Greek mathematicians) adjusted common speech
to precise reasoning, an extraordinarily delicate and improba-
ble achievement. They analyzed motor and visual operations
of great complexity. They made these operations correspond
to grammatical and linguistic properties. Blind seers, they
trusted in speech to guide them through space. And space
itself, as the centuries passed, became an always richer and
more surprising creation; it developed along with thought,
which was acquiring more mastery over itself, and placing
more confidence in the power of reason and in the initial sub-
tlety which had provided it with such incomparable tools:
definitions, axioms, lemmata, theorems, problems, porisms,
and the rest.

—Paul Valéry, *The Intellectual Crisis.*

It would have been possible to put together a book on spirals
that had not a single picture of a spiral; for if you insist on it,
a spiral may be given an axiomatic formulation. It would have
been possible to put together a book on spirals that did not
display a single line of mathematics, for the spiral also lives in
our visually imaginative but pre-axiomatic lives. The first kind
of book might receive the approbation of those mathematicians
who over the past century or so have come to shun the visual as
totally redundant at best and at worst as an a impure source of
logical errors. The second might be welcomed by artists, anthro-
pologists, semioticians for whom mathematics is a bewildering
territory to be avoided.

All the mathematicians that have been included in the historical supplements have, on their part, inserted figures into their
manuscripts. I felt that without figures in considerable profu-
sion this book would have been incomplete and impoverished. I
would assert that a mathematician who is blind and can know
only an axiomatic or a prosthetic spiral, blind seer though he
may be, can know only a shadow. And I find it a profound mystery how the two, the formulaic and the visual, live with each
other and give each other strength. Theories of cognition and
brain physiology may ultimately dispel the mystery. At the moment, the simple knowledge that they do is a great satisfaction

to me, and is part of the excitement that I am able to draw from mathematics.

In condemning a dry-as-dust tendency toward purely symbolic statements, and in advocating a balanced partnership between the visual and the verbal/symbolic, I am well aware of a contemporary threat that now comes (paradoxically) from the visual sector. The computer and computer graphics have in the past generation made and continue to make enormous strides. The ability of computer graphics to create a fictitious reality, a reality of the imagination, is easily as great as that of verbal/symbolic material.

The perception of the threat from the visual quarter sometimes takes the form of the complaint that reading will soon be forgotten, replaced by what is claimed to be a higher, more efficient mode of communication; replaced by the hieroglyph, the icon, the interactive nonlinear text, the terminal and the videocassette. If that should happen, the world of communicated ideas will spiral, retrogressing into an advanced technological barbarism in which people will have little common basis for conducting elevated discourse within traditional languages. It will know little of the strength and beauties of language, spoken or written. Shakespeare, the King James Bible, the Gettysburg Address will be uninterpretable, and even the aesthetics of abstract symbolic formulations will have moved to another plane.

There is growing evidence that we may be moving into such a phase of history in which communication is increasingly visual. The imperialism of the visual derives from the direct and strong impact of computer graphics. The tendency of our students to think only in terms of colored overlays in polywindowed displays, the tendency of our playwrights to substitute motion and kinematic sensation for language, the decline of poetry and recitation, the condemnation of old-fashioned oratory as flatulent and suspicious, are all independent indications of a transition that has been several centuries in the making. Our future teachers of mathematics may become impresarios of hardware and software instead of playwrights, composers or actors, with the more creative roles in the total mathematical production preempted by a small cadre of mathematical élite who work in conjunction with a large group of people skilled in visuals.

We may inquire as to the extent to which this visual component is truly visual in the primitive, nonsymbolic sense that characterizes, say, the precise gymnastics of a squirrel leaping from branch to branch. For beneath each display, each production in computer graphics, lies a massive infra-structure of symbols, organized like a meta-onion, and which, if unbalanced, puts forth imperious claims to precedence and then overwhelms the image. In the days of Cicero there was an argument, not unlike the recent "two-culture argument" of C. P. Snow and F. R. Leavis, as to which was the more important: the word or the deed. We may soon be debating the hyperword versus the hyperimage and then wake up to find they have coalesced into one.

The computer and computer graphics have clearly widened old and opened up new creative sources of the mathematical spirit. A retrospective show of spirals in the year 2500 would amaze us. One hopes that both the visual and the symbolic will endure in their individual aspects as well as in fruitful partnership.

Bibliography

[Abramowitz and Stegun 1964] M. Abramowitz and I. A. Stegun (eds.), *Handbook of Mathematical Functions*, NBS Applied Math. Ser., vol. 55, Washington, D.C.: U.S. Government Printing Office, 1964.

[Aczel and Dhombres 1989] J. Aczel and J. Dhombres, *Functional Equations in Several Variables*. Cambridge: University Press, 1989.

[Aho et al. 1974] Aho, Hopcroft and Ullman, *The Design and Analysis of Computer Algorithms*. Reading: Addison-Wesley, 1974.

[Alberts et al. 1983] B. Alberts et al., *The Molecular Biology of the Cell*, 1983.

[Anderhub 1941] J. H. Anderhub, "Aus den Papieren eines reisenden Kaufmannes," in: *Joco-Seria*. Wiesbaden: Kalle-Werke, 1941.

[Archimedes 1912] Archimedes, *On Spirals*. pp. 151–188, in: *The Works of Archimedes* (T. L. Heath, ed.). New York: Dover. A reprint of the Cambridge University Press edition of 1912.

[Archimedes 1913] Archimedes, *Peri Elikon*, pp. 1–121, in: *Opera Omnia*, vol. II (J. L. Heiberg, ed.). Leipzig: B. G. Teubner, 1913.

[Arnold 1990] V. I. Arnold, *Huygens and Barrow, Newton and Hooke: Pioneers in Mathematical Analysis and Catastrophe Theory from Evolvents to Quasicrystals*. Basel: Birkhäuser, 1990.

[Arnold 1973]. *Ordinary Differential Equations*. Cambridge: MIT Press, 1973.

[Artin 1931] E. Artin, *Einführung in die Theorie der Gammafunktion.* Leipzig, 1931.

[Baass 1984] K. G. Baass, "The use of clothoid templates in highway design," *Transportation Forum* **1** (1984), 47–52.

[Bailey 1989] D. F. Bailey, "Function Iteration," *Mathematics Magazine*, June, 1989.

[Barnsley 1986] M. E. Barnsley, "Making Chaotic Systems to Order," in: Barnsley and Demko [1986].

[Barnsley 1988]. M. E. Barnsley, *Fractals Everywhere.* Boston: Academic Press, 1988.

[Barnsley and Demko 1986] M. E. Barnsley and S. G. Demko, *Chaotic Dynamics and Fractals.* Orlando: Academic Press, 1986.

[Barrow 1860] Isaac Barrow (1630–1677), *The Mathematical Works of Isaac Barrow,* D. D. (Whewell edition), Cambridge, 1860.

[Beardon 1991] Alan F. Beardon, *Iteration of Rational Functions,* New York: Springer-Verlag, 1991.

[Belisle 1989] Claude Belisle, "Windings of random walks," *Annals of Probability* **17** (1989), 1377–1402.

[Bergé and Pomeau 1983] P. Bergé and Vidal Pomeau, *Order within Chaos.* Wiley, Hermann, 1983.

[Bernoulli (Jacobus) 1967] Jacobus (Jakob) Bernoulli (1655–1705), *Opera,* facsimile ed., vol. I, pp. 442–443; 497–502. Bruxelles: 1967.

[Bernoulli (Johannis) 1968] Johannis (Jean, Johann) Bernoulli (1667–1748), *Opera Omnia,* facsimile ed., vol. I. Hildesheim: Georg Olms, 1968.

[Borodin and Munro 1975] A. Borodin and Ian Munro, *The Computational Complexity of Algebraic and Numeric Problems.* Elsevier, 1975.

[Borwein and Borwein 1990] J. Borwein and P. Borwein, *A Dictionary of Real Numbers.* Pacific Grove: Wadsworth and Brooks, 1990.

[Bourgin and Renz 1989] R. D. Bourgin and P. Renz, "Shortest paths in simply connected regions in R^2," *Adv. in Math.* **76** (1989), 260–295.

[Bulmer-Thomas] Ivor Bulmer-Thomas, "Theodorus," in: *Dictionary of Scientific Biography 1970–80.*

[Campbell and Meyer 1979] S. L. Campbell and C. D. Meyer, Jr., *Generalized Inverses of Linear Transformations.* London: Pitman, 1979.

[Caswell 1989] Hal Caswell, *Matrix Population Models.* Sunderland, Mass.: Sinauer Associates, 1989.

[Cesàro 1886] E. Cesàro (1859–1906), "Les lignes barycentriques," *Nouv. Annales de Math.* **5** (1886).

[Cody et al. 1970] W. J. Cody, K. A. Paciorek and H. C. Thacher, Jr., "Chebyshev approximations for Dawson's integral," *Math. Comp.* **24** (1970), 171–178.

[Cook 1979] Theodore A. Cook, *The Curves of Life: being an account of spirals . . . and their application to growth in nature, to science, and to art.* London: Constable, 1914. Reprint: New York: Dover, 1979.

[Coutsias and Kazarinoff 1987] Evangeloss A. Coutsias and Nicholas D. Kazarinoff, "Disorder, renormalizability, theta functions and Cornu spirals," *Physica* **2D** (1987), 295–310.

[Cvitanovic 1984] P. Cvitanovic (ed.), *Universality in Chaos.* Bristol: Hilger, 1984.

[Dahmen et al. 1989] W. Dahmen, M. Gasca and C. A. Micchelli (eds.), *Computation of Curves and Surfaces.* Dordrecht: Kluwer Academic, 1989.

[Davis 1959] Philip J. Davis, "Leonhard Euler's integral: A historical profile of the gamma function," *Amer. Math. Monthly* **66** (1959), 849–869.

[Davis 1963] Philip J. Davis, *Interpolation and Approximation.* New York: Blaisdell, 1963.

[Davis 1974a] Philip J. Davis, *The Schwarz Function and Its Applications.* Carus Mathematical Monograph No. 17. Washington: Mathematical Association of America, 1974.

[Davis 1974b] Philip J. Davis, "Visual geometry, computer graphics, and theorems of perceived type," in: *Proc. of Symposia in Applied Mathematics* **20** (1974). Providence, R.I.: American Mathematical Society.

[Davis 1979] Philip J. Davis, *Circulant Matrices.* New York: Wiley, 1979.

[Davis 1985] Philip J. Davis, "What do I know: A study of mathematical self-awareness," *College Mathematics Journal* **16** (1985), 22–41.

[Davis 1988] Philip J. Davis, *Thomas Gray: Philosopher Cat*, Orlando: Harcourt Brace Jovanovich, 1988.

[Davis 1989] Philip J. Davis, *The Thread*, Orlando: Harcourt Brace Jovanovich, 1989.

[Davis 1991] Philip J. Davis, "Visual theorems," in: *Educational Studies in Mathematics.* Dordrecht: Kluwer Academic, 1991.

[Davis to appear] Philip J. Davis, "The Raised Eyebrows of Mathematics," to appear.

[Davis and Rabinowitz 1984] Philip J. Davis and Philip Rabinowitz, *Methods of Numerical Integration.* 2nd ed. Orlando: Academic Press, 1984.

[Dawson 1898] H. G. Dawson, "On the numerical value of $\int_0^h e^{x^2}\,dx$," *Proc. London Math. Soc.* (1) **29** (1898), 519–522.

[de Bruijn and Szekeres 1955] N. G. de Bruijn and G. Szekeres, "On some exponential and polar representatives of matrices," *Nieuw Arch. Wisk.* **3** (1955), 20–32.

[Dekking and Mendès-France 1981] F. Dekking and M. Mendès-France, "Uniform distribution modulo one: a geometrical viewpoint," *J. Reine Angew. Math.* **329** (1981), 143–153.

[Demetriou and Powell 1990] I. C. Demetriou and M. J. D. Powell, "The minimum sum of squares of change to univariate data that gives convexity," Department of Appl. Math. and Theor. Phys., University of Cambridge, Report 1990/NA 3.

[Deshouillers 1989] Jean-Marc Deshouillers, "Geometric Aspect of Weyl Sums," in Devaney and Keen [1989].

[Devaney 1989] Robert L. Devaney, *Introduction to Chaotic Dynamical Systems*, 2nd ed., Redwood City: Addison-Wesley, 1989.

[Devaney and Keen 1989] Robert L. Devaney and Linda Keen (eds.), "Chaos and fractals: The mathematics behind the computer graphics," *Proc. of Symposia in Appl. Math.* Providence, R.I.: Amererican Mathematical Society, 1989.

[Dijksterhuis, 1956] E. J. Dijksterhuis, *Archimedes*, Copenhagen: Munksgaard, 1956. Princeton: University Press, 1987.

[Fellmann 1985] Emil A. Fellmann, "Die Spirale in der Mathematik," in: Hartmann and Mislin [1985].

[Fellmann 1986] Emil A. Fellmann, "Zur Geschichte der Klothoïde," *Österreichisches Symposium zur Geschichte der Mathematik.* Neuhofen: November, 1986.

[Fischer and Smith 1985] P. Fischer and W. R. Smith (eds.), *Chaos, Fractals, and Dynamics.* New York: Dekker, 1985.

[Foley et al. 1989] T. Foley, T. Goodman and K. Unsworth, "An algorithm for shape preserving parametric interpolating curves with C^2 continuity," pp. 249–260, in: *Mathematical Methods in Computer Aided Geometrical Design* (T. Lyche and L. Schumaker, eds.). Cambridge: Academic Press, 1989.

[Fontanella 1989] F. Fontanella, "Shape-preserving interpolation," pp. 183–214, in: Dahmen et al. [1989].

[Ford 1986] Joseph Ford, "Chaos," in: Barsley and Demko [1986].

[Frazer 1908] J. Frazer, "A new visual illusion of direction," *British Journal of Psychology.* Cambridge, 1908.

[Freud 1971] G. Freud, *Orthogonal Polynomials.* New York: Pergamon Press, 1971.

[Gantmacher 1960] F. R. Gantmacher, *The Theory of Matrices*, New York: Chelsea, 1960.

[Gautschi 1969] W. Gautschi, "Algorithm 363 – Complex error function," *Comm. ACM* **12** (1969), 635.

[Gautschi 1990]. W. Gautschi, "Computational aspects of orthogonal polynomials," in: *Orthogonal Polynomials – Theory and Practice* (P. Nevai, ed.). NATO ASI Series, Series C: Mathematical and Physical Sciences **294** (1990), 181–216. Dordrecht: Kluwer.

[Gautschi 1991] W. Gautschi, "A class of slowly convergent series and their summation by Gaussian quadrature," *Math. Comp.* **57** (1991).

[Gautschi to appear] W. Gautschi, "On a slowly convergent series occurring in plate contact problems," to appear.

[Gautschi and Milovanović 1985] W. Gautschi and G. V. Milovanović, "Gaussian quadrature involving Einstein and Fermi functions with an application to summation of series," *Math. Comp.* **44** (1985), 177–190.

[Gehring 1978] F. W. Gehring, "Spirals and the universal Teichmüller space," *Acta Mathematica* (Uppsala) **141** (1978), 99–113.

[Goldstein 1977] Herman H. Goldstein, *A History of Numerical Analysis from the 16th through the 18th Century.* New York: Springer-Verlag, 1977.

[Golub and Welsch 1969] G. Golub and J. H. Welsch, "Calculation of Gauss quadrature rules," *Math. Comp.* **23** (1969), 221–230. Loose microfiche suppl. A1–A10.

[Gould and Katz 1975] Stephen J. Gould and Michael Katz, "Disruption of ideal geometry in the growth of receptaculitids: a natural experiment in theoretical morphology," *Paleobiology* **1** (1975), 1–20.

[Grabiner 1990] Judith Grabiner, "Was Newton's calculus a dead end? Maclaurin and the Scottish connection," presentation, Mathematical Association of America National Meeting, Ohio State University, Columbus, Ohio, August 8, 1990.

[Gregory et al. 1955] Gregory, Stuart and Walker, *Phil. Trans.* (1955).

[Grenander et al. 1990] Ulf Grenander, Y. S. Chow and D. M. Keenan, *Hands: A Pattern Theoretic Study of Biological Shapes.* New York: Springer-Verlag, 1990.

[Grünbaum and Shephard 1989] Branko Grünbaum and G. C. Shephard, *Tilings and Patterns.* New York: W. H. Freeman, 1989.

[Grünbaum and Shephard 1979] Branko Grünbaum and G. C. Shephard, "Spiral tilings and versatiles," *Mathematics Teaching* (UK) **88** (1979), 50–51.

[Guy 1988] Richard Guy, "The strong law of small numbers," *Amer. Math. Monthly* **95** (1988), 697–712.

[Guy 1990] Richard Guy, "The second strong law of small numbers," *Math. Mag.* **63** (1990), 320.

[Hale and Koçak 1991] Jack Hale and Hüsseyin Koçak, *Differential Equations: An Introduction to Dynamics and Bifurcations.* New York: Springer-Verlag, 1991.

[Halmos 1956] Paul R. Halmos, *Lectures on Ergodic Theory.* New York: Chelsea, 1956.

[Halmos 1990] Paul R. Halmos, "Has progress in mathematics slowed down?" *Amer. Math. Monthly* **97** (1990), 561–588.

[Hammelm et al. 1987] S. M. Hammelm, J.A. Yorke and C. Grebogi, "Do numerical orbits of chaotic dynamical processes represent true orbits?" *J. of Complexity* **3** (1987), 136–145.

[Hao 1984] Bai-Lin Hao, *Chaos*. World Scientific, 1984.

[Hardy and Wright 1960] G. H. Hardy and E. M. Wright, *An Introduction to the Theory of Numbers*. Oxford: University Press, 1960.

[Hartmann and Mislin 1985] Hans Hartmann and Hans Mislin (eds.), *Die Spirale im menschlichen Leben und in der Natur: eine interdisziplinäre Schau*. Basel: Birkhäuser, 1985. (Ausstellung im Museum für Gestaltung, Gewerbemuseum, Basel.)

[Hegstrom and Kondepudi 1990] Hegstrom and Kondepudi, "The handedness of the universe," *Scientific American*, January 1990, 98–105.

[Henrici 1984] P. Henrici, *Applied and Computational Complex Analysis*, vol. I. New York: Wiley, 1984.

[Higgins, J. R. 1985]. "On cardinal expansions," *Bull. Amer. Math. Soc.* **12** (1985), 45–89.

[Hlawka and Binder 1986] E. Hlawka and Christa Binder, "Über die Entwicklung der Theorie der Gleichverteilung in den Jahren 1909 bis 1916," *Archive for History of Science*, 1986.

[Hlawka 1980] E. Hlawka, "Gleichverteilung und Quadratwurzelschnecke," *Monatshefte für Math.* **89** (1980), 19–44.

[Hlawka 1990] E. Hlawka, *Selecta*. Berlin: Springer-Verlag, 1990.

[Hlawka et al. 1981] E. Hlawka, F. Firneis and P. Zinterhof, *Zahlentheoretische Methoden in der numerischen Mathematik*. Wien: R. Oldenbourg, 1981.

[Holden 1986] A. V. Holden (ed.), *Chaos*. Manchester: University Press, 1986.

[Huntley 1970] H. Huntley, *The Divine Proportion*. New York: Dover, 1970.

[Iserles 1988] A. Iserles, Two papers on stability and dynamics of ODE solvers. DAMTP, Cambridge University, 1988/ NA1 & NA5.

[Kabotie 1987] F. Kabotie, *Designs from the Ancient Mimbreños*. Flagstaff, Arizona: Northland Press, 1987.

[Kahane 1985] J.-P. Kahane, *Some Random Series of Functions*, 2nd ed., Chapter 10. Cambridge: University Press, 1985.

[Kairies 1978] H.-H. Kairies, "Convexity in the theory of the gamma function," pp. 49–62, in: *Proceedings of the First International Conference on General Inequalities* (E. F. Beckenbach, ed.). Basel: Birkhäuser, 1978.

[Katz 1988] N. M. Katz, "Gauss sums, Kloosterman sums, and monodromy groups," *Annals of Mathematics Studies*, No. 116. Princeton: University Press, 1988.

[Knorr 1975] W. R. Knorr, *The Evolution of the Euclidean Elements*. Dordrecht/Boston: Reidel, 1975.

[Ko 1986] K. Ko, "Applying techniques of discrete complexity theory to numerical computation," in: *Studies in Complexity Theory* (R. Book, ed.). New York: Wiley, 1986.

[Koçak 1986] Hüseyin Koçak, *Differential and Difference Equations through Computer Experiments*. New York: Springer-Verlag, 1986.

[Kolmogoroff 1949] A. N. Kolmogoroff, "Wiener spirals and some other interesting curves in Hilbert space," *Dokl. Akad. Nauk. SSSR* **26**, 115–118.

[Körner 1988] T. W. Körner, *Fourier Analysis*. Cambridge: University Press, 1988.

[Kostelich and Yorke 1990] E. J. Kostelich and J. A. Yorke, "Noise reduction: Finding the simplest dynamical system consistent with the data," *Physica* D **41** (1990), 183–196.

[Krull 1948, 1949] Wolfgang Krull, "Bemerkungen zur Differenzengleichung $g(x+1) - g(x) = \Phi(x)$," I, II. *Mathematische Nachrichten* **1** Nov./Dez. 1948, 365–376; **2**, März/Apr. 1949, 251–262.

[Kuczma et al. 1968] M. Kuczma, B. Choczewski and R. Ger, *Functional Equations in a Single Variable*. Warsaw: Polish Scientific Publishers, 1968.

[Kuczma et al. 1989] M. Kuczma, B. Choczewski and R. Ger, *Iterative Functional Equations*. Cambridge: University Press, 1989.

[Kuipers and Niederreiter 1974] L. Kuipers and H. Niederreiter, *Uniform Distribution of Sequences*. New York: Wiley, 1974.

[Lakatos 1976] I. Lakatos, *Proofs and Refutations*. Cambridge: University Press, 1976.

[Lakshmikantham and Trigiante 1988] V. Lakshmikantham and D. Trigiante, *Theory of Difference Equations*. Boston: Academic Press, 1988.

[Lancaster 1969] Peter Lancaster, *Theory of Matrices*. New York: Academic Press, 1969.

[Lasalle 1986] J. P. Lasalle, *The Stability and Control of Discrete Processes*. New York: Springer-Verlag, 1986.

[Lasota and Mackey 1986] Andrej Lasota and M. C. Mackey, *Probabilistic Properties of Deterministic Systems*. Cambridge: University Press, 1986.

[Laugwitz and Rodewald 1987] D. Laugwitz and B. Rodewald, "A simple characterization of the gamma function," *Amer. Math. Monthly* **94** (1987), 534–536.

[Lawrence 1972] J. Dennis Lawrence, *A Catalog of Special Plane Curves.* New York: Dover, 1972.

[Leader 1991] Jeffery J. Leader, "The generalized Theodorus iteration," Dissertation. Division of Applied Mathematics, Brown University, July, 1989. UMI Dissertation Information Service, Ann Arbor, 1991. No. 9101795.

[Lichtenberg and Lieberman 1983] A. Lichtenberg and M. Lieberman, *Regular and Stochastic Motion.* Springer-Verlag, New York, 1983.

[Loria 1902] Gino Loria, *Spezielle algebraische und transcendente ebene Kurven.* Leipzig: B. G. Teubner, 1902. Especially vol. 2.

[Lord and Wilson 1984] E. Lord and C. Wilson, *The Mathematical Description of Shape and Form.* New York: Wiley, 1984.

[Loxton and van der Poorten 1977] J. H. Loxton and A. J. van der Poorten, "A class of hypertranscendental functions," *Aequationes Math.* **16** (1977), 93–106.

[Mandelbrot 1977] B. Mandelbrot, *The Fractal Geometry of Nature.* New York: W. H. Freeman, 1977.

[McMahon and Bonner 1983] T. A. McMahon and J. T. Bonner, *On Size and Life.* New York: Scientific American Books, 1983.

[Meek and Walton 1989] D. S. Meek and D. J. Walton, "The use of Cornu spirals in drawing planar curves of controlled curvature," *J. Comput. Appl. Math.* **25** (1989), 69–78.

[Mendès-France 1982] M. Mendès-France, "Paper folding, space filling curves and Rudin-Shapiro sequences," in: *Contemporary Mathematics.* Vol. 9, *Papers in Algebra, Analysis and Statistics.* Providence, R.I.: American Mathematical Society, 1982.

[Melzak 1983] Z. A. Melzak, *Bypasses: A Simple Approach to Complexity.* New York: Wiley, 1983.

[Meschkowski 1959] Herbert Meschkowski, *Differenzengleichungen.* Göttingen: Vandenhoeck & Ruprecht, 1959.

[McCabe 1976] R. McCabe, "Theodorus' irrationality proof," *Mathematics Magazine* **49** (1976), 201–203.

[Milne-Thomson 1933] L. M. Milne-Thomson, *The Calculus of Finite Differences.* London: Macmillan, 1933.

[Neiss 1966] W. Neiss, *Praxis Math.* **8** (1966), 241–3.

[Niederreiter 1978] H. Niederreiter, "Quasi-Monte-Carlo methods and pseudo-random numbers," *Bull. Amer. Math. Soc.* **84** (1978), 957–1041.

[Niven 1961] Ivan Niven, "Uniform distribution of sequences of integers," *Trans. Amer. Math. Soc.* **98** (1961), 52–61.

[Nörlund 1924] N. E. Nörlund, *Vorlesungen über Differenzenrechnung.* Berlin: Springer-Verlag, 1924.

[Nörlund 1929] N. E. Nörlund, *Leçons sur les équations linéaires aux différences finies.* Paris: Gauthier Villars, 1929.

[Ormell 1990] Christopher Ormell, "The end of the defensive era in mathematics," in: *The Mathematical Revolution Inspired by Computing* (Johnson, J. J. and M. J. Loomes, eds.). Oxford: University Press, 1990.

[Pappus of Alexandria 1933] Pappus of Alexandria, *Pappus d'Alexandrie. La Collection Mathématique.* Oeuvre traduite pour la première fois du grec en français par Paul Ver Eecke. Paris, 1933.

[Parlett 1980] B. N. Parlett, *The Symmetric Eigenvalue Problem.* Englewood Cliffs: Prentice Hall, 1980.

[Pauly-Wissowa 1894] Pauly-Wissowa, *Real-Encyclopaedie der classischen Alterthumswissenschaft* (A. Pauly et al., eds.). Stuttgart, 1894 and later.

[Pedoe 1976] Dan Pedoe, *Geometry and the Visual Arts.* New York: Dover Publications, 1976.

[Penrose 1989] Roger Penrose, *The Emperor's New Mind.* Oxford: University Press, 1989.

[Phillips 1981] George M. Phillips, "Archimedes, the numerical analyst," *Amer. Math. Monthly* **88** (1981), 165–169.

[Phillips 1990] George Phillips, Personal communication; February, 1990.

[Pickover 1990] C. A. Pickover, *Computers, Pattern, Chaos, and Beauty.* New York: St. Martin's Press, 1990.

[Pietronero and Tosatti 1986] L. Pietronero and E. Tosatti, *Fractals in Physics.* Amsterdam: North Holland, 1986.

[Plato 1892] Plato, *Theaetetus* (translated by Benjamin Jovett). London: Macmillan, 1892, p. 147.

[Pólya and Szegő 1979] G. Pólya and G. Szegő, *Problems and Theorems in Analysis.* Berlin: Springer-Verlag, 1979.

[Rainville 1967] E. D. Rainville, *Special Functions.* New York: Macmillan, 1967.

[Raup 1966] D. M. Raup, "Geometric analysis of shell coiling," *Journal Paleontology* **40** (1966), 1178–1190.

[Richert 1991] Norman Richert, "Hypocycloids, continued fractions, and distribution modulo one," *Amer. Math. Monthly* **98** (1991), 133–139.

[Rota 1990] G.-C. Rota, "Mathematics and philosophy: The story of a misunderstanding," Cambridge, Mass.: Department of Mathematics, MIT, 1990.

[Rota et al. 1988] G.-C. Rota, D. H. Sharp and R. Sokolowski, "Syntax, semantics, and the problem of the identity of mathematical objects," *Philosophy of Science* **55** (1988), 376–386.

[Rotman in press] Brian Rotman, *Ad Infinitum: The Ghost in Turing's Machine. An essay in corporeal semiotics,* in press.

[Ripley and Sutherland 1990] B. D. Ripley and A. I. Sutherland, "Finding spiral structures in images of galaxies," Report, Dept. of Statistics, University of Strathclyde, Glasgow, Scotland, 1990.

[Rubel 1989] Lee A. Rubel, "A survey of transcendentally transcendental functions," *Amer. Math. Monthly* **96** (1989), 777–788.

[Ruelle 1989] David Ruelle, *Chaotic Evolution and Strange Attractors.* Cambridge: University Press, 1989.

[Sagher 1988] Yoram Sagher, "What Pythagoras could have done," *Amer. Math. Monthly* **95** (1988), 117.

[Sandefur 1990] J. T. Sandefur, *Discrete Dynamical Systems: Theory and Applications.* Oxford: University Press, 1990.

[Schmalz et al. 1984] T. G. Schmalz, G. E. Hite and D. J. Klein, "Compact self-avoiding circuits on two-dimensional lattices," *J. Phys.* A **17** (1984), 445–453.

[Schmidt 1970] W. Schmidt, "Simultaneous approximation to algebraic numbers by rationals," *Acta Arith.* **125** (1970), 189–201.

[Schuster 1988] Heinz G. Schuster, *Deterministic Chaos.* Weinheim: VCH, 1988.

[Schwenk 1976] Theodor Schwenk, *Sensitive Chaos.* New York: Schocken, 1976.

[Senechal and Taylor 1990] Marjorie Senechal and Jean Taylor, "Quasicrystals: The view from Les Houches," *Mathematical Intelligencer*, Spring 1990, 54–64.

[Serra 1982] J. Serra, *Image Analysis and Mathematical Morphology.* London: Academic Press, 1982.

[Spalt 1981] D. D. Spalt, *Vom Mythos der Mathematischen Vernunft,* Darmstadt: Wissenschaftliche Buchgesellschaft Darmstadt, 1981.

[Spanier and Oldham 1987] J. Spanier and K. B. Oldham, *An Atlas of Functions.* New York: Hemisphere Publishing Corp., 1987.

[Sprows 1989] David J. Sprows, "Irrationals and the fundamental theorem of arithmetic," *Amer. Math. Monthly* **96** (1989), 732.

[Stenger 1971] Frank Stenger, "Whittaker's cardinal function in retrospect," *Math. Comp.* **25** (1971), 141–154.

[Stenger 1981] Frank Stenger, "Numerical methods based on Whittaker cardinal or sinc function," *SIAM Review* **23** (1981), 165–224.

[Stevens 1973] Peter Stevens, "Space, architecture and biology," *Syst. Zool.* **22** (1973), 405–408.

[Stevens 1974] Peter Stevens, *Patterns in Nature.* Boston: Little, Brown, 1974.

[Stiefel and Bettis 1969] E. Stiefel and D. G. Bettis, "Stabilization of Cowell's method," *Numer. Math.* **13** (1969), 154–175.

[Stix and Abbott] Margaret Stix and R. T. Abbott, *The Shell: Five Hundred Million Years of Inspired Design.* New York: Harry Abrams.

[Stuart 1967] J. T. Stuart, "Hydrodynamic stability of fluid flows," Inaugural Lecture. London: Imperial College, 1967.

[Sumner 1968] Lloyd Sumner, *Computer Art and Human Response.* Charlottesville, Va., 1968.

[Sylvester 1908] James Joseph Sylvester, *The Collected Mathematical Papers.* Cambridge: University Press, 1908.

[Talbot 1927] Arthur N. Talbot, *The Railway Transition Spiral.* New York: McGraw Hill, 1927.

[Teuffel 1958] E. Teuffel, "Eine Eigenschaft der Quadratwurzelschnecke," *Math. Phys. Semesterberichte* **6** (1958), 148–152.

[Thompson 1917] D'Arcy Wentworth Thompson, *On Growth and Form.* Cambridge: University Press, 1917 (and later editions).

[Torricelli 1955] Evangelista Torricelli (1608–1647), *De infinitis spiralibus.* Translated, edited and commented upon in Italian by Ettore Carruccio. Parallel columns in Latin and Italian. Pisa: Domus Galilaeana, 1955. The original manuscript is circa 1645.

[Tragesser 1984] R. S. Tragesser, *Husserl and Realism in Logic and Mathematics.* Cambridge: University Press, 1984.

[Tricomi 1954] F. G. Tricomi, *Funzioni Ipergeometriche Confluenti.* Roma: Edizioni Cremonese, 1954.

[Turnbull 1939] H. W. James Gregory Turnbull, *Tercentenary Memorial Volume.* London: Bell, 1939.

[van der Waerden 1963] B. L. van der Waerden, *Science Awakening,* pp. 141–146. New York: Wiley, 1963.

[Van Dyke 1982] M. Van Dyke, *An Album of Fluid Motion.* Stanford: Parabolic Press, 1982.

[van Strien 1988] S. van Strien, "Smooth dynamics on the interval," in: *New Directions in Dynamical Systems* (Bedford, T. and J. Swift, eds.). London Math. Soc. Lecture Notes **127**. Cambridge: University Press, 1988.

[Varga 1962] R. S. Varga, *Matrix Iterative Analysis.* Englewood Cliffs: Prentice Hall, 1962.

[de Vargas y Aguierre 1908] J. de Vargas y Aguierre, *Catalógo General de Curvas.* Madrid, 1908.

[Vitale 1975] R. A. Vitale, "Representation of a crinkled arc," *Proc. Amer. Math. Soc.* **52** (1975), 303–304.

[Voderberg 1936, 1937] H. Voderberg, *"Zur Zerlegung der Umgebung eines ebenen Bereiches in Kongruente,"* *Jahresb. der deutschen Mathematiker Verein.* **46** (1936), 229–231; **47** (1937), 159–160.

[von Neumann and Schoenberg 1941] J. von Neumann and I. J. Schoenberg, "Fourier integrals and metric geometry," *Trans. Amer. Math. Soc.* **50** (1941), 226–251.

[von Seggern 1990] D. H. von Seggern, *CRC Handbook of Mathematical Curves and Surfaces.* Boca Raton: CRC Press, 1990.

[Walton and Meek 1989] D. J. Walton and D. S. Meek, "Computer-aided design for horizontal alignment," *J. of Transportation Eng.* **115** (1989), 411–424.

[Washburn and Crowe 1988] D. K. Washburn and D. W. Crowe, *Symmetries of Culture: Theory and Practice of Plane Pattern Analysis*. Seattle: University of Washington Press, 1988.

[Waterhouse 1986] W. C. Waterhouse, "Why square roots are irrational," *Amer. Math. Monthly* **93** (1986), 213–214.

[Widder 1941] D. V. Widder, *The Laplace Transform*. Princeton: University Press, 1941.

[Wimp 1984] Jet Wimp, *Computation with Recurrence Relations*. Boston: Pitman, 1984.

[Yates 1947] Robert C. Yates, *A Handbook on Curves and Their Properties*. Ann Arbor: J. W. Edwards, 1947.

Notes

[1] I have known many mathematicians who would assert: "N'existe pour moi que ce qui me passione; cette phrase trace exactement mes limites" (Julien Green, *Journal*, p. 352).

[2] Hlawka [1980]. For the Quadratwurzelschnecke as inspiration for a problem in Diophantine approximation, see [Hlawka 1990, p. 435].

[3] This is the first place in these lectures where the ideogram ... (dot, dot, dot) occurs in a mathematical sense. I have always believed in giving a finitist interpretation to the so-called infinite insofar as possible. Now the finite seems totally comprehensible and as such has occasioned far less philosophical discussion than the infinite. This is extremely deceptive: I admit that whenever I try to think deeply about the finite, the more it dissolves into incomprehensibility.

In this connection, I should like to advertise a recent book of considerable significance by Brian Rotman [in press]. Rotman's work has "the aim of de-writing the mathematical infinite; that is, replacing the implicit theism of the 'endless' as this occurs in all infinitistic interpretations of the sequence $0, 1, 2, \ldots$ by a semiotic conception of number tied to the physically realizable and the subjectively feasible."

[4] Hardy and Wright [1960]; Hlawka [1980; 1990]; McCabe [1976]; van der Waerden [1963], among others.

The genesis of these talks having been a certain passage in Plato's Theaetetus, it would be amiss not to mention that the general context in which the passage occurs is a discussion of what is knowledge and, more particularly, what is abstraction:

> Suppose that a person were to ask about some very trivial and obvious thing – for example: what is clay? And suppose that we were to reply that there is the clay of potters, there is a clay of oven-makers, there is a clay of brick-makers; would not the answer be ridiculous?"

> —Plato, *Theaetetus*, 147.
> See Historical Supplement B.

This is the Problem of Universals, whose discussion, beginning with Plato and Aristotle, has never really ceased. This is the problem of the particular and the general and of the relationship between the two.

What interests me personally is not so much an account along the lines of epistemology, but historical studies of the interplay between the particular and the general in the development of mathematics. An interesting study could be made of the dynamics of the process of generalization or, to put it slightly differently, of ontogeny and phylogeny in the mathematical field. The evolution of the number concept or of the function concept would serve as instances for a more general description. There are many excellent particular studies available.

Artistically: The writer's problem is "how to strike the balance between the uncommon and the ordinary so as on the one hand to give interest, on the other, to give reality" (Thomas Hardy). To what extent does this balance play a role in mathematics?

Returning to the specific mathematical example in Plato's passage, general theorems covering the case-by-case presentations of Theodorus are now standardly given in books on the elementary theory of numbers. For example: if a and n are positive integers, then the nth root of a is either irrational or it is an integer. If the latter, a is the nth power of an integer.

[5] See [van der Waerden 1964, p. 141]. Readers might enjoy *Thomas Gray, Philosopher Cat* [Davis 1988], a fantasy written by the author of these lectures, placed at the University of Cambridge, and involving Anderhub's "solution."

One person attracted by my fantasy (while he was lecturing on *The Theaetetus*) was Prof. Günter Patzig of the Philosophy Seminar of Georg-August University, Göttingen. On November 1, 1990, he wrote to me that Anderhub (1894–1946) was "a businessman attached to the Kalle Company in Wiesbaden-Bieberich, which produced the famous 'Cellophan' foils which were used for many purposes, especially in the household." In 1913 Anderhub enrolled in the Gymnasium in Laubach, where he read the classics.

Another person so attracted was Dr. Hayo Ahlburg of Alicante, Spain, who on March 23, 1991, wrote that he had made a bit of a hobby of the Quadratwurzelschnecke. He provided me with the following additional references which I now pass along. Unfortunately, I have not had the opportunity to check them out.

Hans Haverman, Problem 789, *J. Recr. Math.,* **11** (4) (1978–9), 301. Solution: Duane Allen, *J. Recr. Math.* **13** (4) (1980–1), 300–303. *J. Recr. Math.* **12** (4), 310. Shmuel Avital, *Crux Mathematicorum*, vol. 11(2), 1985, p.50. Solution: David Singmaster, *Crux Mathematicorum* **12** (7) (1986), 182–184. Hugo Steinhaus, *One Hundred Problems in Elementary Mathematics*, New York: Dover, 1979, p. 14 and p. 69.

[6] The Euler–Maclaurin formula expresses the difference between the value of an integral and its value as computed approximately by the trapezoidal rule. The difference is expressed as an asymptotic series. See, e.g., [Davis and Rabinowitz 1984, p. 106–111.

The formula appears to have deen discovered independently and published by Euler in 1738 and by Maclaurin in 1737. For an interesting historical discussion, see Goldstine, Sec. 2.6.

[7] Fejér's Theorem: Let x_k be a sequence of real numbers, and let $dx_k = x_{k+1} - x_k$. Let $dx_k \to 0$ monotonically and let $k dx_k \to \infty$ as $k \to \infty$. Then the sequence x_k mod 1 is equidistributed in $[0,1]$.

[8] Equidistribution of θ_n in the sense of Weyl means, roughly, that every subinterval of $0 \le i \le 2\pi$ contains, asymptotically, a percentage of the angles θ_n in proportion to the length of the subinterval. Integer equidistribution in the sense of Niven means that when the angles θ_n are rounded down to the nearest integer, all residue classes are visited in asymptotically equal proportions. See [Niederreiter 1978]; [Kuipers and Niederreiter 1974]; [Niven 1961]; [Hlawka and Binder 1986]. For equidistribution of Gauss sums over finite fields, see [Katz 1988]. Equidistribution (*Gleichverteilung* in German) is a deterministic and not a probabilistic theory.

Within chaos theory, equidistribution is a simple case of an "invariant measure" (i.e., the uniform measure).

[9] At this point a word may be in order regarding the relationship of the author of these lectures to the dedicatee.

It is in the nature of publishing that authors receive a good deal of unsolicited mail. This mail variously praises, blames, asks for information, for clarifications, provides the author with what the correspondent thinks is valuable information and insights, supplies a list of errata, and so forth. An author who produces popular material, or even, as the French put it, *haute vulgarisations*, is rather more liable to receive this kind of communication. Over the years, I have received and dealt with my fair share. Generally speaking, the correspondence terminates after one exchange.

In January 1986, I received a letter by courier post from N. M. Rothschild and Sons, the City, London. Lord Rothschild had interested himself in the distribution of the prime numbers, especially in their probabilistic aspect, and asked me to clarify a point in one of my books. From this initial inquiry, our correspondence grew over four years to a warm friendship. The correspondence dealt almost exclusively with the distribution of prime numbers, but when we met in Cambridge on numerous occasions, our conversation ranged widely over personalities and ideas.

I was no expert on number theory. But I had had three graduate courses in analytic number theory and this proved more than sufficient to shape my answers. The very last material we discussed was the so-called Kac–Erdős theorem, which puts the prime number theorem in a probabilistic context. This theorem can be found in Marc Kac's very beautiful book *Statistical Independence in Probability, Analysis and Number Theory* (MAA/John Wiley, 1959). Kac presented the 1955 Hedrick Lectures, and his book derives from the lectures.

Victor Rothschild was a distinguished biologist and a Fellow of the Royal Society. He was also a considerable public figure. At the time of his death, I had been working for some while on the material presented here. I think he would have been pleased by this dedication.

[10] See, e.g., [Vitale 1975; Kahane 1985]. von Neumann and Schoenberg [1941] call $P(t)$ a helix if $|P(t) - P(s)| = \theta(t - s)$. This jibes with the three-dimensional helix. The function θ is called the screw function and is characterized in their paper.

In the Hilbert space of Brownian trajectories, the variance condition $W(t)$ describes a helix. That is, the distance between two points depends only on the difference of the time parameters.

[11] There is also a substantial and popular "morphological" literature about spirals. See, e.g., [Cook 1979; Huntley 1970; Lord and Wilson 1989; McMahon and Bonner 1983; Schwenk 1976; Stevens 1973, 1974; Stix and Abbott]. The pioneering work of D'Arcy Wentworth Thompson [1917, (esp. Chap. 6)] should also be cited here. More technical citations would include: [S. J. Gould and M. J. Katz 1975], [P. S. Stevens 1973], [Alberts et al. 1983, pp. 564–570].

Some of the popular literature displays what might be called "Φ-morpho-mysticism." (Φ = the golden number = $(\frac{1}{2})(1 + \sqrt{5})$. There is naturally a spira aurea, a golden spiral linked to Φ, whose polar equation is $r = \exp(\gamma\theta)$, where $\gamma = (2/\pi)\log(\Phi) = .30634896 \, 2.5$

Some years ago, π- and e-mysticism were more frequent than they seem to be now. Example: isn't it remarkable that π, the circle number, appears in the normal probability distribution – thus linking the lengths of people's lives with the eternal circle! Φ-mysticism seems to be more persistent.

The doctrine that "all is number" is known as Pythagoreanism. Neo-Pythagoreanism has both inspired and afflicted numerous investigators from cosmologist Arthur Eddington to some contemporary students of chaos. It can move rapidly into theosophy and the occult.

For a demystification of the Fibonacci sequence, and hence the golden number, in floral growth, Stevens [1974, p. 166] has this to say:

> It simply grows its stalks or florets in succession around the apex of the stem so that each fits the gaps of the others. The plant is not in love with the Fibonacci series; it does not even count its stalks; it just puts out stalks where they will have the most room. All the beauty and all the mathematics are a natural by-products of a simple system of growth interacting with its spatial environment.

[12] See [Hartmann and Mislin 1985]. This book contains twenty-four "popular" articles ranging widely over all aspects of spirals; from spirals as art forms and as Jungian symbols to spirals in economics and in molecular biology.

[13] Archimedes: 287 (?)–212 B.C. See [Dijksterhuis 1956, Chapter VIII]. Other results on spirals from classical Greek mathematics include the spirals on cylinders, cones and spheres. (See [Pappus of Alexandria 1933, Chapter VIII, 57 and Chapter IV, 53; in the translation of Ver Eecke, pp. 878 and 201].) In these lectures, I will tend to use the word "spiral" independently of the dimension.

[14] The quadrature of the spiral was approached by Archimedes in this way: he approximates the area of a spiral sector by a finite number of circular sectors (whose area he knows). He then allows the number of circular sectors to grow so as to exhaust the area of the spiral sector. His final answer depends on the fact that (in modern notation) he can evaluate $(1^2 + 2^2 + \cdots + n^2)$ and hence find $\lim(1/n^3)(1^2 + 2^2 + \cdots + n^2)$ as $n \to \infty$. See [Dijksterhuis 1956, p. 277] and Historical Supplement C.

The problem of finding a neat formula for the total area swept out by the (discrete) spiral of Theodorus would have stumped Archimedes, since it calls for an evaluation of

$$S_n = \sum_{k=1}^{k=n}(1/\sqrt{k}).$$

X ?   This requires knowledge of the Euler–Maclaurin theorem as in (1.1).

****

While in today's mathematical scene research workers are completely
aware of the unsolved problems in their field, we tend to forget that
the ancient world must also have been replete with unsolved problems.
What were they?

One has heard, of course, of the three famous unsolved problems
of classical Greek mathematics: the ruler-and-compass quadrature of
the circle, duplication of the cube and the trisection of the angle.
(Unsolved in the sense of having to come to grips with self-imposed
limitations.) These were not discussed adequately until the nineteenth
century.

But what were some of the other unsolved problems? We can only
conjecture. The classical ideal of presentation (and much emulated
today) was that of the highly organized, highly polished, backwardly
arranged document, with all evidence expunged of the struggle to
arrive at it. And then the transmission of classical mathematics would
have tended to filter out all that was inconclusive. In the mediaeval
period, things began to relax a bit and authors occasionally would
say: "I did thus and so and it didn't work. Then I did so and so and
I still didn't get anywhere. So there we are."

[15] On Konon, see [Pauly-Wissowa 1894, XI, pp. 1338-1340].

A letter from Prof. Gerald Toomer, dated July 7, 1989:

"In answer to your query, the story that Konon (of Samos, not
Alexandria, although he did live in Alexandria for a while, where he
made himself endeared to the monarch by claiming to have discovered
a lock of the Queens hair – the Coma Berenices – there is a famous
poem on the subject by Callimachus, translated into Latin by Catullus) was responsible for a theorem on the spiral is found in Pappus,
Collection Book IV, 30, (Hultsch, p. 234), where he (Pappus) says
that Konon propounded the theorem on the spiral, but Archimedes
proved it."

Whether this is really true we have no means of knowing. Pappus,
writing in the fourth century A.D. did have access to many mathematical works, now lost. But he is far from reliable, and in this case
it seems probable to me that he had exactly the same information
that we have, namely, Archimedes' book on spirals, where in the introduction, Archimedes, addressing Dositheus, says that he had sent
these theorems to Konon long ago, and that Konon would have found
proofs of them, but unfortunately died before he could do so.

From Archimedes, then, I conclude that he sent Konon the enun-
ciation of the propositions, without the proofs (a nasty habit of some

later mathematicians also), that Dositheus inherited these from Konon, and after Konon's death, wrote to Archimedes asking for the proofs, which Archimedes supplied in the extant book, and that Pappus has once again made a muddle."

[16] These sad thoughts raise the question: To what extent does mathematics as we know it depend upon the individual? Or is mathematics just sitting out there waiting for whatever talented shovel comes along to scoop it up?

In the humanistics, I think one would assert that both the *Symphony in G-minor*, and *The Importance of Being Ernest* depended very much on Mozart and on Oscar Wilde as individuals. In mathematics and in science generally, I think one is less likely to make such assertions. The number of simultaneous discoveries is great, and there is a tendency in any case to smooth out over the years the individual approach towards a common point of view and understanding.

[17] The soothsayers were right after all. Berenice was ultimately poisoned, probably with the connivance of her husband.

[18] After these lectures were delivered, a well-known research mathematician, Q, came up to me and asked : What is a spiral of Bernoulli? I was mildly shocked that he did not know. For the sake of completeness then, here are the polar equations for the spirals of Archimedes and of Bernoulli:

$$\text{Archimedes: } r = k\theta, k = \text{constant.}$$

$$\text{Bernoulli: } r = \exp(k\theta), k = \text{constant.}$$

Apparently Q was quite able to pursue a productive research career without this knowledge. This experience raises the question: is there a core of material that every educated mathematician ought to know? Given that the amount of mathematical information is far beyond any single person's grasp, what balance should be reached between the concrete and the abstract, the pure and the applied, the new and the old, the continuous and the discrete, mathematics for the millions as against mathematics for the mathematical élite.

Having answered this question to your satisfaction, answer the next question: Is there a core of material that every educated person should know? I bring this up in view of the battle, currently raging in academic circles, regarding Eurocentrist versus multiculturalist versus particularist curricula.

[19] For example, spiral vortices in the boundary layer of a rotating disc are quite accurately Bernoullian. See, e.g., [Gregory, Stuart and Walker 1955].

I cannot resist appending a quote from [McMahon and Bonner 1983], a lovely book on morphology.

> Each time the Nautilus outgrows part of its old living chamber, it creates a new empty chamber about 6.3% larger than its predecessor,.... The result is that in the course of building eighteen chambers needed to bring this spiral full circle, the size of the chambers triples.

Now think of that! And the nautilus has always been modelled by a spiral of Bernoulli, and not by a spiral of Theodorus. (Fig. 5.) Thus the criticality and universality of the number 17 is established, a fact which no mathematician of imagination has doubted ever since Gauss showed how the regular 17-gon can be constructed with ruler and compass!

As an additional biological complication: there are non-Bernoullian snails. See [Raup 1966].

[20] Torricelli "rectified" the logarithmic spiral, i.e., found its arc length, in what, apparently, was the first rectification of the arc of a curve in the history of mathematics. This possibility had been despaired of by numerous mathematicians including Descartes. I use the term rectification in the strong sense, i.e., he constructed by elementary geometry a straight line segment whose length equals that of the spiral arc. He also showed that as the spiral winds around its asymptotic point infinitely often, its arc length remains finite.

James Gregory was concerned with tangents to the spiral. See [Turnbull 1939] for allusions in Gregory's correspondence to this question.

Jakob Bernoulli showed that the evolute, the pedal curve, the caustic by reflection, and the caustic by refraction of a logarithmic spiral are also logarithmic spirals. Bernoulli also considered spiral planetary orbits arising from various gravitational laws.

The geometrical literature of the late nineteenth century often represents spirals (and curves in general) in terms of "pedal coordinates": $r = f(p)$, where r = length of radius vector, p = length of perpendicular from the origin to the tangent to the curve at the point in question. The notion of pedal coordinates goes back at least to Maclaurin in 1718. Here are some equations in pedal coordinates.

In general: $p = r^2/\sqrt{r^2 + (dr/d\theta)^2}$.

The spiral of Bernoulli: $p = ar, a = \text{const.}$

The spiral of Archimedes: $p = r^2/\sqrt{a^2 + r^2}$

The Cotes' spiral, i.e., the path of a particle moving under the inverse cube law of attraction: $p^{-2} = (ar^{-2}) + b$.

Pedal coordinates are akin to the familiar support function in convex body theory, which, in the two-dimensional case would be $p(\theta)$.

Then, of course, there are the representations in terms of "natural" or "intrinsic" coordinates of arc length s and curvature k. In these coordinates, sk = constant is the spiral of Bernoulli, while s/k = constant is the spiral of Cornu. (See fig. 8.)

[21] It has been pointed out many times that the spiral on Bernoulli's monument seems to be Archimedean rather than Bernoullian. This piece of mathematical irony calls for an explanation, and there has developed a considerable literature that offers a number of explanations. (Cf. the Theodorus controversy as another instance of a peculiar kind of extra-mathematical controversy that mathematics can engender.) For an introduction to the literature of this strange chapter in the history of mathematics, see [Fellmann 1985].

The inscription on the monument reads "resurrectionem piorum hic praestolatur": here he awaits the resurrection of the pious.

Here are Bernoulli's words. I give them in the original Latin in fervent hope of a resurrection of the study of classic languages:

Cum autem ob proprietatem tam singularem tamque admirabilem mire mihi placeat spira haec mirabilis, sic ut ejus contemplatione satiari vix queam, cogitavi illam ad varias res symbolice repraesentandas noninconcinne adhiberi posse. Quoniam enim semper sibi similem et eandem spiram gignit, utcumque volvatur, radiet; hinc poterit esse vel sobolis parentibus per omnia similis Emblema; Simillima Filia Matri: vel (si rem aeternae veritatis Fidei mysteriis accomodare non est prohibitum) ipsius aeternae generationis Filii, qui Patris velut Imago. et ab illo ut Lumen a Lumine emanans eidem $o\mu o\sigma\iota o\sigma$ existit, qualiscumque adumbratio. Aut, si mavis, quia Curva nostra mirabilis in ipsa mutatione semper sibi constantissime manet similis et numero eadem, poterit esse vel fortitudinis et constantiae in adversitatibus; vel etiam Carnis nostrae post varias alterationes, et tandem ipsam quoque mortem ejusdem numero resurecturae symbolum; adeo quidem, ut si Archimedem imitandi hodiernum consuetudo obtineret, libenter Spiram hanc tumulo meo juberem incidi cum Epigraphae : Eadem numero mutata resurget.

See [Loria 1902, vol. II, p. 67], or [Bernoulli (Jacobus) 1967, *Opera*, vol. I, p. 502].

As he says, he is ordering this motto in imitation of Archimedes, who had a sphere and its circumscribed right cylinder carved on his stone. Speaking of lapidary inscriptions, particularly of the Swiss variety, Alexander Ostrowski (1893–1986), late Professor of Mathematics at the University of Basel, must have been impressed by Bernoulli's memorial, which was not too far from his office. In his house in Montagnola, near Lugano, in a wall adjacent to a central fireplace, he caused the fundamental properties of valuation theory (for which he was responsible) to be carved in concrete. One wonders what the subsequent inhabitants made of these occult characters.

[22] For an interesting recent discussion of particle orbits under central force laws of arbitrary power p, placed in a historical context, see [Arnold 1990, Appendix I].

Working in the complex plane, if the orbit is designated by $z(t)$, then the equation of motion is

$$z'' = -cz|z|^p, c > 0. \qquad (*)$$

The value $p = -3$ corresponds to the usual law of gravity. If z'' is discretized by replacing it by $(z_{n+2} - 2z_{n+1} + z_n)/h^2, t_{n+1} - t_n = h$, then (*) can be translated into an iteration of Theodorus type with

$$A = \begin{pmatrix} 2 & -1 \\ 1 & 0 \end{pmatrix}, \quad B = -ch^2 \begin{pmatrix} 0 & 1 \\ 0 & 0 \end{pmatrix},$$

$v_n = \begin{pmatrix} z_n \\ z_{n+1} \end{pmatrix}$, and where $\| v_n \|$ is the quasi $-$ norm $\| z_n \|^3$.

See [Davis 1974, Chap. 9] for the equations of the spirals of Bernoulli and Archimedes in conjugate coordinates, i.e., z and $\text{conj}(z)$. In these coordinates, the spiral of Bernoulli can be regarded as the "ith power of a circle": $(i = \sqrt{-1})$! Further "i powerings" produce the cycle: circle, spiral, real axis, spiral, circle. This is my only contribution to the literature of self-renewal, and if I were given to mathematical mysticism, I could surely cash in on it.

Self-renewal, self-reproduction, self-similarity, self-reference, the macrocosm in the microcosm, and so forth, whether in myth (the phoenix), in religion (the Osiris story, the Easter story), or in logic (the Cretan liar); whether in graphics (fractals) or in analysis (eigenvectors); whether leading to paradox or to revelation; all have been elevated by some authors to the status of a Grand Principle of the Universe. It may be apropos to recall the words of Lord Melbourne that no one ever did anything very foolish except in the name of a strong principle.

[23] See, e.g., [Loria 1902, vol. II, p. 146], or [Sylvester 1908, vol. II, p. 639], where it will be found embedded in a general theory of the

evolutes of a circle. The spiral of Norwich is the evolute of the evolute of a circle. Sylvester lists a half dozen "remarkable" properties of this spiral. The polar equation of this spiral is

$$\theta + c = \sqrt{((r - a)/a)} - 2 \arccos{(\sqrt{a/r})},$$

where a and c are constants. See also: Historical Supplement F.

[24] The parametric equation of a "standard" clothoid is $z(s) = \int_0^s \exp(is^2)ds$.

With regard to the beauty of visual mathematical objects, R. L. Devaney [1990, p. 6] reports that the Mandelbrot set "has been called the most complicated yet the most beautiful object in mathematics." See also [Pickover 1990] for graphical objects generated by iteration. Aesthetic judgements of this sort – even group judgements – are notoriously time-dependent. Take a look, say, at the 1920 winners of the Miss America Contest.

On various attempts to reduce the beauty of proportion to a mathematical formula, see [Pedoe 1976, Chap. 4].

[25] There are at least thirty special mathematical spirals listed in [de Vargas y Aguirre 1908]. To top this list, Prof. Trevor Stuart has kindly pointed out to me that The Spirals are a pop group. So the auditory dimension has been heard from! For a number of pictures of spirals in 3-d, see [Pickover 1990; von Seggern 1990].

[26] If an object "winds around," is it a spiral? In this connection, cf. Spitzer's law for the winding around of a random walk and its extension by Belisle: let $z(t)$ be a standard two-dimensional Brownian motion starting at a point $z(0) \neq 0$, and let $\theta(t) =$ the total continuous angle wound around the origin by z up to time t. Then as $t \to \infty$, $2\theta(t)/\log(t)$ converges in distribution to c, where c is a Cauchy random variable. See [Belisle 1989].

[27] For recent difficulties in defining a quasicrystal, see Senechal and Taylor 1990].

[28] See [Grünbaum and Shephard 1979, Chapter 9].

[29]Note the fundamental paradox: the more that can be asserted about chaos, the less it merits that designation. This was already noted by Abraham de Moivre: "[True chance] can neither be defined nor understood: nor can any proposition concerning it be either affirmed or denied, excepting this one: 'That it is a mere word.'" (A. de Moivre, *Doctrine of Chances*, London, 1756. Cited by Lorraine Daston in her *Classical Probability in the Enlightenment*. Daston remarks further that the "classical probabilists ... strenuously denied both the subjective and objective existence of real chance," p. 11.)

See Joseph Ford's article in [Barnsley and Demko 1986]. A couple of "Joseph Fordisms" may be quoted: "Chaos means deterministic randomness." "Chaos emerges as a mystery – a Gödel-type mystery which only a god can understand."

This appeal to the divine reminds me of Leibniz' (1702) characterization of imaginary numbers: "The imaginary numbers are a subtle and wonderful haven [*Zuflucht*] of the Divine Spirit; they are almost an amphibian between Being and Not-Being." (*Leibnizens mathematische Schriften*, K. I. Gerhardt (ed.). Asher, Berlin, 1849–63, vol. 5, p. 357.)

It is no longer thought necessary to invoke the deity when dealing with complex numbers. As with this concept, future scientists may axiomatize chaos to a fare-thee-well or embed it (or embalm it) in firmer, nonchaotic conceptualizations. But will that suffice to exorcise the ontological questions? As Aristotle implied, true chaos is unknowable, because knowledge equals form. Ford's remark is easily justified.

PLATONISM ? [handwritten marginal note]

[30] Mathematical definitions limit the universe of discourse and in this way create a restricted universe in which it is found possible to pursue the subject along certain traditional lines. The two universes must be carefully distinguished.

William James has a beautiful description of this process (formulated for philosophy but equally applicable to mathematics), which I recommend be read aloud at the beginning of every course on Mathematical Modelling:

A young graduate student said that

> he had always taken for granted that when you entered a philosophic classroom you had to open relations with a universe entirely distinct from the one you left behind you in the street. The two were supposed, he said, to have so little to do with each other, that you could not possibly occupy your mind with them at the same time. The world of concrete personal experiences to which the street belongs is multitudinous beyond imagination, tangled, muddied, painful and perplexed. The world to which your philosophy professor introduces you is simple, clean and noble. The contradictions of real life are absent from it. Its architecture is classic. Principles of reason trace its outlines, logical necessities cement its parts. Purity and dignity are what it most expresses. It is a kind of marble temple shining on a hill.
>
> In point of fact it is far less an account of this actual world than a clear construction built upon it, a classic sanctuary in which the rationalist fancy may take refuge from the intolerably confused and gothic character which mere facts present. It is no explanation of our concrete universe,

it is another thing altogether, a substitute for it, a remedy, a way of escape.

—William James, *Pragmatism*: Lecture I.

Goethe had put it in poetry:

Wer will was Lebendig's erkennen und beschreiben,
sucht erst den Geist herauszutreiben,
dann hat er die Teile in seiner Hand,
fehlt leider nur das geistige Band.

(*Faust I*, lines 1936–9)

To analyze a living creature
One first drives out its spirit nature
The simple parts then lie in hand
Too bad they lack the living strand.

[31] One older definition of a spiral is that it is a curve whose polar form is

$$r = f(\theta)/(\theta + g(\theta)),$$

where f and g are rational (complex) trigonometric functions of θ. This would include the exponentials. (G. Fouret, *Nouv. Ann. Math.* *2e Sér.* **19**, 1880. See [Loria 1902, vol. II, p. 53].)

F. W. Gehring [1978] finds the following definition appropriate to his inquiries: an open arc in the complex plane is a spiral from z_1 onto z_2 if it has the representation

$$z_t = (z_1 - z_2)r(t)\exp(it) + z_2, t \in (1, \infty),$$

where $r(t)$ is positive, continuous and with $\lim_{t\to 0} r(t) = 1, \lim_{t\to\infty} r(t) = 0$

[32] For example: the pointwise product of the discrete spiral of Theodorus and the marigold is (up to $O(n^{-1})$) the 8-ray spiral whose points lie on eight rays inclined consecutively at 45^0.

[33] The Lie theory of spirals was already considered by Sophus Lie and Felix Klein, and the spiral of Bernoulli placed therein. A simple spiral, say the spiral of Archimedes, considered as an individual, is not preserved under any of the symmetry groups that are traditionally used to classify designs. (See, e.g., [Washburn and Crowe 1988].) The clothoid, of course, is preserved under rotations through π as is the case with the "full" spiral of Archimedes. (See fig. 52.) Many of the spirals generated by the Theodorus iterations exhibit "almost" or "asymptotic" symmetries.

***

It is probably the case that apart from applications, the principal

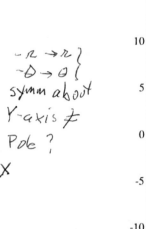

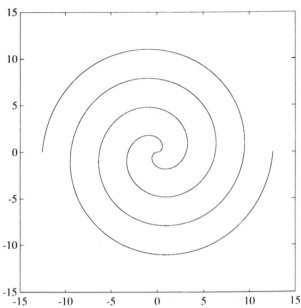

(handwritten margin note: $-\mathcal{R} \to \mathcal{R}$ / $-\theta \to \theta$ symm about Y-axis $\neq$ Pole ? X)

Figure 52:   The "full" spiral of Archimedes exhibiting rotational symmetry.

(handwritten margin note: P.225)

mathematical interest today in spirals is that they are one of the canonical figures arising in the eigenanalysis of the critical points of solutions of ODEs. Recall the phase portrait analysis of systems of real autonomous differential equations: $y' = f(y)$, where f is regular at the origin and $f(0) = 0$. If J is the Jacobian of f, and if (in 2-d) the eigenvalues of $J(0)$ are complex conjugate and have negative real part, then 0 is a stable focus or a spiral point, (in German, a *Strudelpunkt*). In the neighborhood of the origin the orbits are approximately affine maps of spirals of Bernoulli. See [Arnold 1973, Chap. 3; see also p. 139, where the spiral orbits in 3-d are divided into three types, depending on the eigenvalue location in the complex plane].

For f nonlinear, there may be a closed curve towards which the orbits spiral. This is a limit cycle. There are analogous results for systems of difference equations.

***

J. Serra sets up four principles to be satisfied by the morphological transformations useful for the purpose of the quantification of a geometric structure. They are (1) compatibility under translation; (2) compatability under change of scale; (3) principle of local structure; (4) semi-continuity. Serra's book stresses convex body theory and Minkowski set operations.

[34] Under a wide subclass of digital (i.e., discrete) linear filters, the marigold will be preserved morphologically. For the definition of the Kronecker product of two matrices, see any good book on matrix theory.

[35] In the context of discrete graphics, differentiating = differencing, and integrating = cumulative summing.

[36] The spiral of Cornu was discovered by Euler. Cesàro dubbed it the clothoid from *klothein*, to spin. Clotho was one of the three Fates. She spun the Thread of Human Destiny. See [Cesàro 1886; also Fellman 1985].

Cesàro's Thread of Destiny is built into highway design (at least in Europe) and constitutes an interesting chapter in the applied theory of spline functions. Highways tend to be straight lines and circular arcs connected up so as to provide C^1 continuity. At constant speeds, this gives rise to a discontinuity in the acceleration vector, and at high speeds, drivers will tend to overcome this by increasing their radius, which means driving along the chord of the arc. What is advocated is a linear variation in the magnitude of the acceleration, and this can be achieved by blending with a clothoidal arc. (Cf. the instrinsic equation that serves to characterize the clothoid: arc length/curvature = constant.)

This aspect of highway design has given rise to a considerable literature of clothoidic splines. (See fig. 53.) See [Baass 1989; Meek and Walton 1989; Talbot 1927; Walton and Meek 1989]; and further references therein.

[37] Isaac Barrow (1630–1677), Lucasian Professor of Mathematics at the University of Cambridge. Vol. I, p. 66 of [Barrow 1860].

[38] As an instance of Peirce's ugly but presumably precise words, I cite 'agapasticism', which, apparently, is a synthesis of the concepts of 'tychism'and 'synechism'. My fifteen-hundred-page dictionary lists none of these three words.

[39] A novelist of my acquaintance once told me: "I've always felt that spirals have about them something slightly erotic, mischievous and not quite proper."

[40] Consider also the aperçu of L. Wittgenstein:

If humans were not in general agreed about the colour of things, if undetermined cases were not exceptional, then our

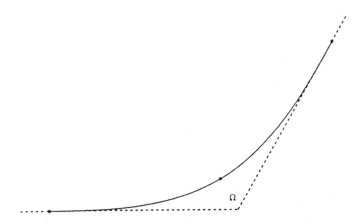

Figure 53:   Clothoid spline: blending two straight lines by means of two clothoid spirals. *After Meek and Walton.*

concepts of colour could not exist. No – our concept would not exist.

> —L. Wittgenstein, *Zettel*, Anscome, G. E. M. and
> G. H. Von Wright, eds., Basil Blackwell, Oxford, 1967.

Or as the cliché of the French high schools puts it: "Ce qui se conçoit bien s'énonce clairement. Les mots pour le dire viennent aisément."

Now, replace the word "colour" in the Wittgenstein quote with the word "spirality." Can we conclude from this that the "spirality" of curves is generally agreed on, but that no satisfactory formalization of this agreement is possible?

[41] David Berlinski has written me ". . . mathematical concepts themselves have something like an intrinsic topology. But what topology and why are certain mathematical concepts stable and others not?"

[42] For a bestiary of strange spirals, see Pickover [1990, particularly Section 12.3]. There are some wonderful color plates of spiral surfaces reproduced in this book.

A definition of a spiral tiling is attempted on p. 515 of Grünbaum and Shephard [1979]. According to a letter from B. Grünbaum to the author dated April 30, 1990, "Even that definition is not really satisfactory. It may well be that the psychological aspect is more important than any mathematical criteria."

This puts us back to the question posed by Socrates in the *Theaetetus*: 'What is clay?' So now it can be told about spirals as a general concept: A 'spiral' is a name for a certain kind of experience and judgement, and is not uniquely or completely reducible to mathematical symbols or to algorithmization. For a discussion of the passage in Plato, see John McDowell, Plato: *Theaetetus*, Clarendon Press, Oxford, 1973, pp. 114–116.

Also relevant to this discussion is the following observation of G.-C. Rota:

> Mathematicians take mischievous pleasure in faking the arbitrariness of definition. In actual fact, no mathematical definition is arbitrary. The theorems of mathematics motivate the definitions as much as the definitions motivate the theorems. A good definition is 'justified' by the theorems one can prove with it, just like the proof of a theorem is 'justified' by appealing to a previously given definition.

There is thus a hidden circularity in formal mathematical definition. The theorems are proved starting with definitions, but the definitions themselves are motivated by the theorems that we have previously decided ought to be in the canon.

But one should not get hung up about introducing definitions. History shows that overconcern with definitions and categories as was evinced, e.g., by the Aristotelians, often leads to a dead end. What interests me is why certain concepts have been made precise and seem to remain stable, while others have not.

I should allow my favorite philosopher to sum it up:

> There is no complete generalization, no total point of view, no all-pervasive unity, but everywhere some residual resistance to verbalization, formulation, and discursification, some genius of reality that escapes from the pressure of the logical finger, that says hands off and claims its privacy, and means to be left to its own life.
>
> —William James, *A Pluralistic Universe.*

For some recent discussions of mathematical ontology, see Rota, Sharp and Sokolowski (phenomenologist) [1988]; Spalt (platonist/formalist) [1981].

[43] On the other hand, spirals may be "seen" where they "should not" be seen. In Frazer's Illusion, a sequence of concentric circular arcs is organized by the eye into a spiral that "isn't really there." See [Frazer 1908], and many books on optical illusions. Frazer's illusory spiral has even been drawn into a discussion of phenomenalist (Husserlian) philosophy of mathematics. See [Tragesser 1984], Chap. 3.

[44] By considering various standard methods of numerical solution, Runge-Kutta, etc., we are led to an entirely different sort of generalization of the (discrete) spiral of Theodorus. See Iserles [1988] for a discussion of the dynamics of this type of situation and its relation to questions of numerical stability. In connection with the numerical integration of differential equations of the harmonic oscillator type, see [Stiefel and Bettis 1969] for the concepts of orbital and phase stability and for algorithms that stabilize Cowell's method.

In accepting computer output as the solution of a differential system, two questions must be faced: (a) truncation error and (b) roundoff error. Question (a) asks: How close do the exact solutions of the difference scheme come to the exact solutions of the differential system? Question (b) asks: In view of computer roundoff, how close does the computed values of the diffence scheme come to the exact solution of the difference scheme?

Toward an answer to (b), we may cite the so-called shadow lemma of Anosov-Bowen, which is for the iteration $x_{n+1} = f(x_n)$. Roughly, it says this: For a restricted class of maps f known as hyperbolic, a true orbit can be found near the computer-produced orbit for arbitrarily long times. See [Hammelm et al. 1987].

While it is of great interest and importance to produce theoretical results such as this, the problem as seen from the "bottom line" of application is somewhat different. The modelling process starts with a real world situation and ends with computer output that, hopefully, leads to usable predictions or insight into that situation. Instead of splitting the modelling process into the mathematical model, the algorithmic surrogate, and the computer output to that algorithm on a specific computer using specific software, one can keep it all together conceptually and ask whether the whole process can be validated as yielding something useful.

In these lectures, of course, we are concerned exclusively with difference equations, so only question (b) is relevant. The reader must keep in mind that the orbits displayed in figures are the computed orbits and must be distinguished from the "true" orbits. Our understanding of the relationship between what is "true" and what is "computable" is currently in an unsettled state.

[45] See also the work of Ripley and Sutherland [1990], who have been setting up a descriminator for galaxies, nebulae, etc., in astrophysics.

This work is conceptually allied to the fundamental work of Ulf Grenander on pattern recognition. Grenander et al. [1990] deal with the question: "Is it possible to mechanize human intuitive understanding of biological pictures that typically exhibit a lot of variability but also possess characteristic structure?"

[46] I was reacting here to an increasing amount of discussion as to what is and what isn't mathematics.

For examples, Paul Halmos [1990, p. 577] calls into question computer proofs: "Appel and Haken do not completely share my religion." The word "religion" may have been used by Halmos in the spirit of irony, but I think it goes beyond that. The Aristotelian spirit set prohibitions against what it called metabasis, i.e., the mixture of modes of thought and operation from one discipline to another. So, in mathematics, there was a tabu against mixing the modes of mathematics and mechanics (motion). These prohibitions were themselves occasioned by a feeling that the universe is so diverse it cannot be unified by one method, and that the integrity and purity of its parts are violated by attempts to do so.

The elevation of the "mind" over the "eye" that occurred in nineteenth- and twentieth-century mathematics, or, as we have seen, the "mind" over the "computer," is, I think, a modern manifestation of the horror of metabasis [Davis 1974; 1991].

Edmund Gibbon remarked that "for the people in the Roman Empire all religions are equally true; for the philosophers all were equally false, and for the magistrates, all equally useful." Mathematics is, in part, a religion in that it is based ultimately on our faith in the meaningfulness, the coherence and the stability of certain kinds of thought processes. Which mathematical "church" you adhere to at a given moment of time depends on whether you feel yourself one of the people, one of the philosophers, or one of the magistrates.

So what is mathematics? This question is hardly discussed in any course in mathematics taken by undergraduates, and there are probably as many answers as there are people who attempt an answer.

For starters: mathematics is the science of quantity and space. This answer might have been satisfactory three or four hundred years ago. Today we might want to amplify it by saying that mathematics is the science and art of deductive and algorithmic structures that concern themselves with quantity, space, pattern and arrangement and of the symbolisms by which this is accomplished. That is a bit more comprehensive, but there may still be special interest groups (semioticians, physicists, non-Cantorian set theorists) who might feel that something has still been left out.

One thing is certain: mathematics is far too important a subject to be left to the mathematicians either for definition, extension or promulgation.

[47] See, e.g, the article on Theodorus by Ivor Bulmer-Thomas in the *Dictionary of Scientific Biography*. Also [Pauly-Wissowa 1894, 2nd. ser. X. pp. 1811–1825].

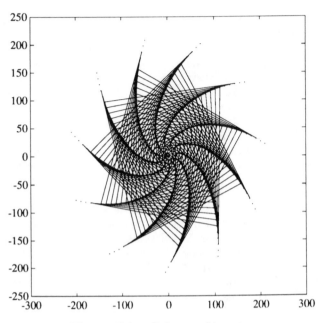

Figure 54:   Spirographic art.

[48] The successive iterates are easily created on a piece of paper with a pencil and a right angled triangle. This is equally true of the "pinwheel spiral." This reminds me of a children's toy called "spirograph," which consists of a number of geared or ratcheted plastic wheels, rings and racks and ovals. By rotating one piece meshed with another and allowing the pencil point inserted in punched holes to trace continuous orbits, "spirographic" designs are created. (They are really families of epi- and hypocycloids.)

Another mechanical toy, perhaps a bit more sophisticated, was the "Magic Designer" or "Hoot-Nanny." Though computer graphics can create things a million times more sophisticated, there is still virtue in working in a minimalist manner and returning to the visual simplicities. In defense of the primitive, it should be pointed out that a landscape, whether natural or mathematical, differs if one walks through it or one drives through it at 70 mph.

Spirography, interpreted in a fairly wide sense, was the basis, a quarter of a century ago, of a certain school of computer art. See, e.g., [Sumner 1968]. (See fig. 54.)

In recent years, phenomenal advances both in hardware and software, resulting in stunning productions, have altered this. Judging from the SIGGRAPH 1990 Show of computer art (See: *Computer Graphics*, Vol. 24, No. 6), trompe-l'œil three-dimensional modeling and the creation of surface texture via fractals and other means dominated the exhibition. The older computer simplicities have been driven off the computer art market. See, e.g., [Pickover 1990, Chapter 13].

Today, the largest outlet for computer art is in commercial graphics. It seems to have "lost the thrill of the ultra-modern that it had twenty years ago, and it doesn't seem to be at the center of the concerns of practising artists." – C. S. Strauss.

[49] See [Davis 1959].

[50] Some of the early developers of the theory and practice of spline functions were I. J. Schoenberg, and J. H. Ahlberg, E. Nielsen, and J. L. Walsh. Spline theory has grown mightily in many directions and has become an absolute staple in many areas of numerical analysis and computer-aided geometrical design. Any graphical design package worth its salt will have a variety of spline programs conveniently on tap.

[51] A possibility: the spiral is of the form $r = f(\theta)$, where $f \epsilon C^2$, and f, f', and f'' are all ≥ 0. Or why not simply say that the curve has positive curvature? (See fig. 55.) This would lead to the second order differential inequality

$$k(f) = (f^2) - ff'' + 2(f')^2 \geq 0.$$

These two conditions are independent: take, e.g., $r = f(\theta) = \exp(\theta^2)$. Then over $0 \leq \theta \leq 1, f, f$, and $f'' \geq 0$, while the curvature changes sign! On the other hand, with $f = \theta^{1/2}, k(f) \geq 0$, but $f, f' \geq 0$ while $f'' < 0$.

Or consider these geometric definitions: a curve will be called spirally (or polar) convex if any sufficiently small open arc is separated from the origin by its chord; or, a curve will be called midpoint spirally convex if for sufficiently small arcs, the "midpoint" of the arc (anglewise, not arclengthwise) is separated from the origin by the chord.

It is interesting to note that both Archimedes and Torricelli, in pre-calculus days, had inserted into their monographs convexity statements for their respective spirals [Torricelli 1955, Par. 5].

[52] I think this function is original with me; but it is so much in the spirit of seventeenth and eighteenth century mathematics that I should not be surprised to learn it was already discussed in those years.

This suggests an interesting question: To what extent can one mathematical age create new mathematics in the spirit of a previous age? What purpose could such a piece of work serve?

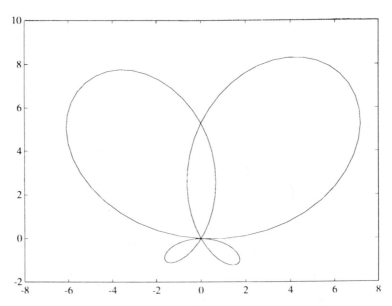

Figure 55:   Spirally convex (second definition).

I once wrote that I. J. Schoenberg's fundamental work in the theory of spline functions looked superficially as though it might have been created in the days of Euler, but that this would be a mistaken view. It would be better, I added, to compare it with Prokofieff's *Classical Symphony*. Schoenberg was a fine musician, and this comment pleased him very much.

The painter Auguste Renoir had this to say:

> It is impossible to repeat in one period what was done in another. The point of view is not the same, any more than are the tools, the ideas, the needs, or the painters' techniques.

In writing mathematical history, one should certainly discuss the point of view, the tools, the ideas, the needs, the techniques of a particular age. However, the desire to explicate a particular piece of work in terms of its ultimate position in today's mathematical corpus is very strong. (As a simple and almost omnipresent aspect of this: mathematics of an older period is often rewritten in contemporary notation so as to be "more comprehensible." Such a translation is not really possible; it asserts that the semiotic content of a mathematical text is time invariant.)

To present the past seen as a justification of the present, or the present seen as the logical and inevitable completion of the past is known as "Whig" history. It is hard to avoid, valuable in its way, but it does not allow one to get "under the skin" of the original creation.

[53] Other possible roads to interpolation: use the asymptotic expansion (1.1), or, perhaps, assuming we know the values at the integers, expand in a series of "cardinal" functions. See, e.g., [Stenger 1981]. As the Cheshire Cat in *Alice in Wonderland* inferred, the right direction to take depends on where you want to be when you get there: a two point boundary problem. Initially, it didn't matter too much with me. Anywhere was O.K.

[54] Nörlund's theory may be conveniently read in [Milne-Thomson 1933, Chapter 8].

It should be observed that the problem of analytic interpolation may be rephrased as the problem of finding the fractional or continuous iterates to a map, in this case, the Theodorus map $z \to z + iz/|z|$. See (2.1).

If one designates a fractional iterate $f^{(t)}(x)$ of the map $f(x)$ by $F(t,x)$, then F satisfies, formally, the functional equation

$$F(s, F(t,x)) = F(s+t, x),$$

for all appropriate s, t, s. This is known in the literature of functional equations as the translation equation. Designate by h "any" function with an inverse $h^{-1} : h(h^{-1}(x)) = x$. Then it is easy to check that formally

$$F(t,x) = h^{-1}(t + h(x))$$

will satisfy the translation equation. If the variables in question are real, then under certain continuity conditions of F, one can prove that there must be a continuous, strictly monotonic $h(x)$ that does the trick. One must now connect h with f:

$$f(x) = f^1(x) = h^{-1}(1 + h(x)).$$

This leads immediately to

$$h(f(x)) = h(x) + 1,$$

which is the so-called Abel's equation. See, e.g., [Aczel and Dhombres 1989, p. 297].

[55] This leads us to a new chapter of interpolation theory known as constrained interpolation. In addition to data matching, we require that the interpolant satisfy certain inequality contraints. This is of great importance in, e.g., computer-aided geometrical design, and a

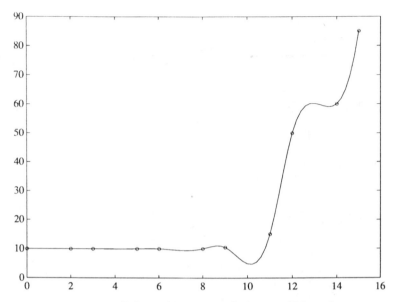

Figure 56:   Cubic spline interpolating to Akima data.

considerable literature both theoretical and algorithmic has grown up in the last generation.

A subproblem to constrained interpolation/approximation is that of "shape-preserving interpolation/approximation." The question is this. Given in, say, R^2, a finite amount of data $(x_k, y_k), k = 0, 1, \ldots, N, x_0 < x_1 < \ldots$. We wish to interpolate or approximate this data from among a family S of explicitly given functions in such a way that the "shape" of the data (interpreted initially in a subjective sense) is preserved. Important additional requirements might be that the interpolant satisfy a certain functional equation, that the resulting numerical algorithms be stable and robust, and of reasonably low computer complexity.

One way to objectify the subjective notion of shape preservation is to require that the interpolant, $s(x)$, extracted from the family S, mimic the monotonicity and/or the convexity character of the data. Thus, setting $h_k = x_{k+1} - x_k, \Delta_k = (y_{k+1} - y_k)/h_k$, one would require that

$$sgn(s'(x)) = sgn(\Delta k) \quad \text{for} \quad x \in [x_k, x_{k+1}],$$

$$sgn(s''(x)) = sgn[\Delta_{k+1} - \Delta_k] \quad \text{for} \quad x \in [x_k, x_{k+1}],$$

$k = 0, 1, \ldots, N - 1$, or both simultaneously.

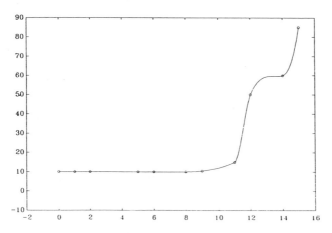

Figure 57:   Cubic spline shape-preserving interpolant. *From:*
Fontanella.

There are some existential surprises in these requirements. Suppose
that S_{nk} designates the class of splines that are piecewise polynomials of degree less than equal n over the mesh $x_0, x_1, \ldots, x_N$ and of
continuity class C^k there. It has been proved by Passow and Roulier
that for every $n > 1$, there is convex data (second difference positive)
for which there can be no convex interpolating spline $s(x) \in S_{n1}$. See
[Fontanella 1989]. (See figs. 56, 57.)

Historically, one of the first shape-preserving transformations is to
replace discrete, equispaced data by its Bernstein polynomial. This
will preserve monotonicity and convexity over the given interval. See
[Davis 1963, pp. 114–115].

For a recent introduction into the literature of shape-preserving interpolation, Fontanella's article is highly recommended. Further material can be found in [Foley et al. 1989]. For convexity preservation
of data via least square methods, see [Demetriou and Powell 1990].

As far as I am aware, the question of shape preservation for data
that is to be organized "spirally" is quite open.

[56] See, e.g., [Artin 1931]. The condition of logarithmic convexity
cannot be replaced by simple convexity. A theorem of H. H. Kaires
tells us that if the constant t is positive and sufficiently small, ($t <
\min \psi'(a)/4\pi^2, \psi(2)/4\pi), \psi = \Gamma'/\Gamma$, then, on the positive real axis, the
function $g(x) = \Gamma(x) \exp(t \sin 2\pi x)$ will be analytic, convex, logarith-
mically convex on $(0, a]$, satisfy the difference equation for the gamma
function, satisfy the functional equation $g(x)g(1 - x) = \pi/\sin(\pi x)$
for the gamma function, and $g(1) = 1$. Yet, obviously, $g \neq \Gamma$, so that

mere convexity gives the first difference equation plenty of leeway. For a concrete example, take $a = 2$. Then $\psi(2) = .42278$ and $\psi'(2) = .64493$, so that $t = .01$ will work.

For this and other characterizations of the gamma function see [Kairies 1978]; also: [Kuczma et al. 1989, Chap. 10, Sect. 4.], and [Laugwitz and Rodewald 1987].

As regards possible functional equations for the Theodorus function (2.7), apart from (2.3) and what may be implied by the development in (2.14)–(2.18), none have yet come forward.

[57] I thought that an identification of $T(a)$ would prove difficult. I had just received a copy of Borwein and Borwein's book *A Dictionary of Real Numbers* [Borwein and Borwein 1990], which lists 100,000 "interesting" real numbers. From the Preface to the dictionary:

> How do we recognize that the number $.93371663\ldots$ is actually $2\log_{10}(e + \pi)/2$? Gauss observed that the number $1.85407467\ldots$ is (essentially) a rational value of an elliptic integral – an observation that was critical in the development of nineteenth century analysis.

If I were lucky, my world constant T would be in the Borwein dictionary and that might give me a clue. It wasn't, so it didn't. Herbert Wilf reports the same experience with regard to some universal constants that have come up in the recent theory of the Josephus problem.

There is a story, possibly apocryphal, that Charles Darwin once recommended that all scientists at some point in their career should put some money on a long shot. When asked what had he done along these lines, Darwin answered that he had played the trombone to a bed of tulips. *The Dictionary of Real Numbers* may be in that category.

$$* * *$$

The theory of computer complexity has opened up new questions (or restimulated some old ideas) with regard to real numbers: given a real number x defined in some way, how long, asymptotically speaking, does it take to compute the first k digits of x? For example, if x is an integer, the nth digit of $\sqrt{x}$ can be computed in $O(n\log n\log\log n)$ time. Can you find an answer that is best possible?

It would be interesting to compare the asymptotic complexity of Gautschi's method against that of Phillips.

See [Borodin and Munro 1975]; [Ko 1986]; [Aho et al. 1974].

$$* * *$$

On the designation of the constant of "Theororus" by the letter T: in conformity with the Greek background of the man, I should have designated it with a (capital theta). On the fun and agony of transliterations of foreign alphabets into English, see [Davis 1989, *The Thread*], where it is described how the transliteration of 'Tschebyscheff' lead me to a few adventures. This book continues to elicit correspondence. Richard Valentin sent me the following clip from "Mathematical Anecdotes" by Stephen G. Krantz, *Mathematical Intelligencer*, Vol. 12, No. 4, Fall 1990, p. 35.

> (Abram S.) Besicovitch (of almost periodic function fame) was a smart man, so he quickly became proficient at English. But it was never perfect. He adhered to the Russian paradigm of never using articles before nouns. One day, during his lecture, the class chuckled at his fractured English. Besicovitch turned to the audience and said 'Gentlemen, there are fifty million Englishmen speak English you speak; there are two hundred million Russians speak English I speak.'
>
> In another lecture series, on approximation theory, he announced 'Zere is no t in ze name Chebyshóv.' Two weeks later he said 'Ve now introduce ze class of T-polynomials because T is ze first letter of ze name Chebyshóv.'

[58] A difficult question. There used to be a debate as to how mathematics might have developed if, say, Archimedes had been in possession of a modern electronic digital computer. The discussion persists today in the form of speculations on possible future developments of mathematics in view of the existence of the computer. See, e.g., [Ormell 1990] and also the volume in which Ormell's article is embedded.

[59] However, it is by no means the case that the simple precedes the complex, chronologically speaking. It may take generations to simplify statements, objects, proofs or whatever.

The spiral of Theodorus, fantasized to be prior to the spiral of Archimedes, and though visually indistinguishable from it, is a much more complicated object, and its continuous version would have been beyond the ability of second-century B.C. Greek mathematics to deal with. This fact could bolster the fantasy that it was discovered early and then dropped.

[60] A very appropriate tack for a St. Andrews mathematician. See a previous note on the Euler–Maclaurin formula.

Judith Grabiner [1990] has written a nice article showing the great importance of Maclaurin's work to continental mathematicians; a fact that has been forgotten (or never realized) by most contemporary mathematicians.

[61] For lovers of long numbers (arithmosophiles, to coin an expression that gentrifies the slightly pejorative "number freaks"),

$$T = 1.8600250792211903072\ldots.$$

[62] Of course, Gautschi (occasionally together with Milovanovic) had been working for some years on the theoretical and the numerical analysis infrastructure of a related problem. Without this infrastructure and the associated software in place, T could not have been computed so accurately and so expeditiously by these means.

What enthralled me particularly was the role that the Gaussian formula for approximate integration played in his method. It took me back to my work at the National Bureau of Standards, in the days of the first generation of digital computers, when Philip Rabinowitz and I were the first to compute Gaussian integration weights and abscissas electronically.

***

That Gautschi should have visited Brown was "pure coincidence." A theory that coincidences are by no means coincidences has been put forward by Persi Diaconis. It is still an open question whether God does or does not shoot craps with the Universe.

The real problem, of course, is how to know all that is "known" in mathematics that is relevant to a specific problem at hand. Handbooks, compendia, the Silver Disc, Mathematica, Sketchpad-type packages, as necessary as they are, do not come close to answering this question. Compilations or computerizations of results may themselves be thought to be part of the algorithmization of mathematics. See [Davis 1985]. For recent opinions on the limitations of the algorithmic mode, see [Penrose 1989, Chap. 10].

[63] For an introduction to the features of Dawson's integral, see [Spanier and Oldham 1987, Chap. 42].

On the device employed by Gautschi see Ostrowski's remark in note 64. This device is an interesting example of what Z. A. Melzak has called the "bypass" or "conjugacy" principle. The second expression is taken from algebra where STS^{-1} is the conjugate of T under S. Speaking generally and vaguely, the bypass device uses S to map a problem into a domain where the transformed problem is easier. Designating the operation of solving by T, one then uses S^{-1} to map the solution back into the original environment.

Melzak [1983] has assembled a wide and interesting collection of problems from all areas of pure and applied mathematics where this principle has been put to good use. The use of logarithms to reduce multiplication to addition is probably the first place where the student of mathematics is introduced to the methodology, and the reader of Melzak's book will be able to augment his examples with many others.

Melzak makes wide claims for the "STS^{-1} principle." But whether this is merely a part of the grammar of mathematical methodology or something deeper is not clear.

[64] Another reminiscence. Gautschi was a Ph.D. student of Alexander Ostrowski at the University of Basel. Ostrowski was not particularly inclined toward philosophy. Science fiction and sensational literature of that sort were more his cup of tea.

I recall a conversation with Ostrowski as we were riding together in a bus at a conference held at the General Motors Research laboratory.

> Philosophy. Ah, yes, philosophy. You know that Boris Pasternak (Soviet Nobelist in Literature, 1890–1960) and I were students together at the University of Marburg. I was in the faculty of science. He was in the faculty of philosophy. Philosophy of mathematics? Well, I'll tell you. In the seventeenth and eighteenth centuries, mathematicians tried to express integrals as sums. In the nineteenth century they began to express sums as integrals. So mathematics goes in spirals. That's all there is to it.

[65] Dawson appears to have been a British school teacher in the 1890s.

[66] A transcendentally transcendental function (or a TT function for short) is one that cannot satisfy an ordinary differential equation with algebraic coefficients. The Hölder–Ostrowski theorem assures us that the gamma function is TT. See the very nice article by Lee Rubel [1989], and further references there; also [Loxton and van der Poorten 1977]. As Rubel wrote me in a letter dated 19 December, 1989, "No one, as far as I can tell, has any general methods for such problems. It's all hunt and peck and the kitchen sink."

Incidentally, the first modern classification of numbers and functions into algebraic, transcendental numbers or functions, etc., is due to James Gregory in his *Vera Quadratura* (Padua, 1667). This book also contains the first modern treatment of convergence and of systems of difference equations. See [Turnbull 1939]; also [Bailey 1989].

[67] For more on the Schroeder function (particularly in the context of the reflection principle for analytic functions), see, e.g., [Davis 1974a, *The Schwarz Function*]. To read about the Schroeder function in the context of chaos and fractals and Julia sets, see [Devaney 1989, Section 3.4], and the article by L. Keen in [Devaney and Keen 1989]. See also [Beardon 1991]. The most extensive treatment of the Schroeder

function and its generalizations can be found in Chaps. 8, 9 and 11 of [Kuczma et al. 1989].

The set of all a on the unit circle for which $a^n \neq 1$ and for which there is a Schroeder function is known as the Siegel set. For a number of sufficient conditions for membership in the Siegel set, see [Kuczma et al. 1989, p. 156].

[68] *Theodora Goes Wild:* a light-hearted comedy starring Irene Dunne, 1936.

[69] See [Caswell 1989], particularly Chaps. 8 and 9. Chapter 9 has some analysis of density-dependent population models. The term $v_n/\|v_n\|$ in our general Theodorus iteration may be interpreted as a population density vector. When the model is of the form $v_{n+1} = h(v_n)Av_n$, where h is a scalar function and A is constant, irreducible, nonnegative and primitive, an ergodic theorem is given.

For the linear Leslie population model in the context of backward population projection and the so-called Drazin inverse, see [Campbell and Meyer 1979, pp. 184–7].

[70] In this case, one should require G to be singular. For the general solution of the linear homogeneous equation $Gv_{n+1} = Hv_n$, expressed in terms of the Drazin inverses of G and H, see [Campell and Meyer 1979, pp. 182–3].

[71] See, e.g., [Barnsley 1988, p. 91], for the random iteration algorithm and accompanying fractal art. The computer implementation of randomness is not random but deterministic (pseudorandom), so one is pretending to be operating here at the boundary between two major mathematical concepts neither of which, in view of recent developments in computational dynamical systems, is any longer clear. As Fred Astaire said in *The Gay Divorcée*, " 'chance' is what the fool calls 'fate'," and not too many people have noticed that this sentence can be read either that fate is foolishly called chance or vice versa.

[72] Notice: we may write the last as $v_{n+1} = A_n v_n$, and therefore almost any sequence of vectors may be produced as the result of a linear time-varying process. All we need is to take $A_n = v_{n+1} \, \text{pinv} \, (v_n)$, assuming that no $v_k = 0$. We have written pinv for the Moore–Penrose generalized inverse.

The necessity of reining in one's generalizations contradicts the old saw about how the essence of mathematics lies in its freedom. I would say that the essence of mathematics lies in the tension between what you will let it do, what you won't let it do, and what you can't let it do.

If one can do "everything" with a certain type of algorithm, then the game changes. The question becomes: With what algorithm of a certain meta-type can one create, for example the Mona Lisa, with the fewest parameters or the fewest bits of information, or the fewest lines of programming? Data compactification of this sort is a form of Ockham's Razor.

For example, the solutions of linear, homogeneous, constant coefficient systems of differential or difference equations are polyexponential functions. (See, e.g., [Arnold 1973, p. 103, 176]. Arnold calls them quasi-polynomials.) The question then becomes what can you do if you limit yourself to the space of such functions. See also [Kostelich and Yorke 1990], who, within the context of noise reduction, ask for the simplest iteration consistent with given data.

This reminds me of one of the late Marc Kac's favorite gags: "With six parameters I can draw an elephant. With seven I can make its tail wag." The whole of mathematics can be regarded as a data compactification language.

[73] Plotting every ninth iterate of the marigold is particularly interesting. For a discussion relevant to morphology of how unorganized discrete spirals can be organized into sub-spirals, see [Gould and Katz 1975].

[74] For example: there appear to be $9, 17, 25$ and 33 "petals" in the first, second, third and fourth rings of the marigold. "Prove" it. How many petals will there be in the fifth ring, not shown in the illustration? If r is the number of the ring, would you conjecture that the number of petals will be $8r + 1$, so that for $r = 5$ the number is 41? Indeed, will there be a fifth ring of petals? Recall also that the figure has been produced by a digital computer whose arithmetic is only an approximation to standard arithmetic.

The reader who is inclined to jump to conclusions on the basis of the first few cases should look at the two articles by Richard Guy [1988; 1990], who has put together an amusing collection of examples. Guy's moral is: "You can't tell by looking." And yet, we are always jumping to conclusions on the basis of what seems to be a finite amount of information.

***

The metaphysics of the standard approach to the calculus of limits strikes me as deficient. Within this theory, the limit of a convergent sequence of numbers is not affected one whit if the first N numbers of the sequence are altered. Therefore, in principle at least, one cannot tell the limit of a sequence by examining any finite number of individual terms. In practice, this is done every day of the year and with considerable success.

[75] It seems hardly necessary to give references in an age when chaos/fractals have exploded like a nova and become buzz words;

when they have been apotheosized and given rise to a minor industry of applied and pseudoapplied mathematics; when they have even penetrated secondary schools, the computer amusement, and the T-shirt markets. But here are a few: [Barnsley 1988; Barnsley and Demko 1989; Beardon 1991; Bergé and Pomeau; Cvitanovic 1984; Devaney 1989; Devaney and Keen 1989; Fischer and Smith 1985; Hale and Koçak 1991; Hao 1984; Holden 1986; Mandelbrot 1977; Pickover 1990; Ruelle 1989; Schuster 1988].

A few words might be in order describing how numerical analysts view chaos/fractals. Iterative methods lie at the very heart of numerical analysis and the development of algorithms for scientific computation. The numerical analyst looks for methods that are stable (convergent), robust (remaining stable over a wide variety of inputs) and economical. A method that is nonconvergent is discarded without further ado. (Too bad; back to the drawing board! On second thought, what was proposed was so reasonable, perhaps we can find a way of modifying, adapting, relaxing the iterations so as to force convergence or to accelerate it.) Convergent methods lead to uninteresting graphics: a set of dots moving towards a limit point. The visual equation is

$$\text{robust} = \text{boring.}$$

The iterations leading to fractals/chaos derive from unstable processes; hence nonconvergent iterations. Bounded orbits, in particular, lead to interesting graphics. It would be a serious mistake to say that the current research interest in fractals/chaos derives from its visual aspect, but it is surely the case that this aspect has enabled the subject to "go public" in a big way.

At first, chaos theoreticians had to relearn or rediscover many things that numerical analysts knew from experience. Then this turned around, and the wide interest in fractals has encouraged numerical analysts to reexamine what happens at the fringes of stability. I am thinking of such investigators as Morris, Gouley, and Mitchell at the University of Dundee. For example, the "leap frog" scheme for $u_t = u_x$ that replaces the derivatives by central differences has been examined outside the stability region described by $(\Delta t)/(\Delta x) < 1$.

The deeper philosophical issue of the application of fractals is really not whether such and such processes are stable/robust, but what physical realities can fractal algorithms mimic? Insofar as fractals exhibit a self-referencing property, this is really the ancient problem of the macrocosm residing in the microcosm.

[76] But see [Mandelbrot 1977], Chapter 10. Texture is also reminiscent of patterns in weaving of fabrics, and this, in turn, reminds us

that one of the first automated digital "computers" was the Jacquard
(1752–1834) loom. So we have spiralled again. See also [Pickover
1990].

Some aspects of texture in equidistribution mod 1 of the sequence
$x_n = (n\theta)$ have been analyzed by Richert in terms of sequence discrepancy. (See Historical Supplement H for this concept.) If θ is rational,
then $(n\theta)$ is periodic. If θ is "close" to a rational number in the sense
that its continued fraction expansion has small partial quotients, then
there is a high discrepancy and the points of the sequence have small
tendency to clump together. On the other hand, if θ is "far" from a
rational, clumping will be observed. If θ is selected as $(\frac{1}{2})(1 + \sqrt{5}) =$
the golden number, or $\theta = \sqrt{2}$, it is in the former category. On the
other hand, for $\theta = \frac{1}{\pi}$, the clumping is strong.

By plotting the points $z_n = \exp(2\pi\theta n i)$ for $n = 0, \ldots, 200$, and connecting them by straight lines – as is common in graphical software
– clumping (or lack of it) is made vivid. As the number of points increases to about 500, and as equidistribution takes over, the clumping
disappears visually. The dynamic progression from initial clumping to
equidistribution as more and more points are plotted is striking. (See
fig. 58.)

The concern with issues of stability/instability in differential and
difference equations in the past hundred years – I should say the
overconcern with these issues – goes beyond matters of theoretical
or practical interest. I think it is related to the rapid advancement
of science and technology and to the great social, psychological and
political instabilities and upheavals in the wake of this advance. This
has given rise to a literature of scientific apocalypticism that has been
rife since the publication of H. G. Wells's *War of the Worlds* almost
a century ago. Popular descriptions of chaos/fractal phenomena are
often couched in apocalyptic language: e.g., "A Siegel disc crumbles,"
"An explosion into chaos", etc. (From R. L. Devaney, *Chaos, Fractals and Dynamics*.) I find it quite understandable that the Soviet
Union, whose stability was maintained for more than a half century
by draconic social measures, should have produced some of the most
brilliant students of mathematical stability.

Stability is theoreticians' substitute for salvation; but

salvation = boring,

for the human heart seems to crave a great measure of excitement.
The lesson that only in limited measure can salvation be found in
theorems is one that is yet to be learned.

[77] One can argue for the recognition of a class of computer-generated
"visual theorems" that might be incorporated into our mathematical

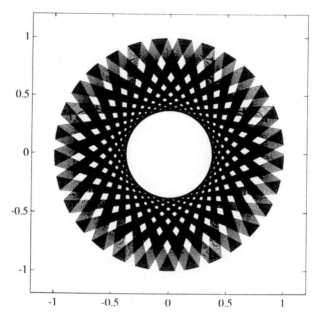

Figure 58:   Exhibiting clumping.

$$z(n) = \exp(2\pi i n\theta), \theta = \tfrac{1}{10}(\sqrt{3} + \sqrt{17} + \sqrt{63}).$$

experience in a way that is rather different from the traditional modality of definition–theorem–proof. It is not even necessary to verbalize, let alone formalize, what one sees as the algorithmic output. See, e.g., [Davis 1974b]. Does this contradict the idea that mathematics is a process of verbal or symbolic communication at both the input and output ends?

When you look at the marigold, what do you see? (See figs. 28, 29.) What I see is a round shape with petallike convolutions distributed radially in a certain manner. But I see much more than this verbal description. I see a totally integrated complicated texture some of whose features I might be able to reduce to words. I call this kind of thing a visual theorem and by this term I mean the passage from the mathematical iteration to the perceived figure grasped and intuited in all its stateable and unstateable visual complexities.

In a letter dated 29 November, 1990, the morphologist Prof. Michael

Katz, of the Bio-architectonics Center, School of Medicine, Case Western Reserve, says that the marigold of Theodorus

> has the flavor of many biological structures. Generally, I would say that these are architectures built from a number of essentially similar units–either many similar molecules or many similar cells or many similar organisms. I once had dealings with a reminiscent set of fossil, Receptaculitids, built of circular and spiral patterns. I would also mention some of the multicellular or colonial protozoans, such as Volvox. Standing back from the pattern, I see that it has a bit of the repetitive, circular patterns that I associate with flower petals, with sunflower seeds, and with the white fluffy stages of dandelions.

Speaking in his persona as an "anecdotalist," Prof. Katz adds that the marigold of Theodorus

> reminds him of an Edward Koren New Yorker cartoon. It looks fluffy and active and busy. It feels like it is complex, but not overwhelmingly so. It is a figure that can involve the human eye – which means that it is one of the endless configurations that can engage naturally and must therefore resonate with the innate structure of the human nervous system.

This visual theorem has permanence (or stability) in that when I run the iteration on my computer, I always get the marigold. With high probability, it is the case that you will get it on your computer. The marigold is distinguishable or classifiable in the sense that if you change the coefficients in the iteration sufficiently, you will get another figure that is recognizably different. The totality of figures that result from all iterations of the type that gave rise to the marigold can be given various taxonomies, one of which is visual, and another would be located within the theory of discrete dynamical systems.

The passage from the iteration to the visual image embodies the elements of definition (the iteration), proof (the actual running of the program on a specific computer), and theorem (the pair: [iteration/computer, visual image]). Now here is surely a nonstandard concept of a theorem that ought to raise a few eyebrows.

Many questions of a traditional mathematical type can be raised about the marigold and its generalizations. Some of these have been raised and some answered in these lectures and in the supplements.

One may also ask in a totally different vein:

(a) In view of the superficial visual connection to the biological world between the marigold and other figures such as spirals, spiders, etc., generated by closely related iterations, can you apply this type of iteration to morphology (or indeed, to any purpose at all)?

(b) Do you think that you can see something in the figure that cannot be described in words? How would that fit in with the old principle of linguistics that "if you can't say it, you can't think it"?

(c) Can this visual theorem contribute to one's emotional life? [When I first hit upon this visual theorem, quite by accident, I experienced a definite sense of elation and delight equal to that experienced when I first encountered certain traditional theorems (e.g., the theorem of Pythagoras or Schur's theorem that any square matrix may be unitarily upper triangularized).] For more on these topics, see [Davis, 1991].

And if visual theorems, why not auditory theorems? One might listen to a spiral, for example, via its discrete Fourier transform. A standard demo of the Mathematica package provides us with the Riemann zeta function as an auditory experience.

[78] When the vectors are complex, I have occasionally found the following simple strategy to be useful visually: display each higher-dimensional vector $[z_1, z_2, \ldots, z_n]$ as a closed polygon $P : (z_1, z_2, \ldots, z_n)$ in 2-d. If the vectors are real, complexify them by pairing components.

[79] The complex variable spirals of Theodorus have the following neat invariance with respect to rotations: let p be an arbitrary complex number and let $q = \Omega p, \Omega = \exp(i\theta), 0 \leq \theta < 2\pi$. Let sp be the (discrete) spiral that is generated by starting at $z = p$, while sq starts at $z = q$. Then $sq = \Omega sp$.

In the higher-dimensional cases, we should require that the unitary transformation Ω commute with A and with B. If Ω is taken as the discrete Fourier matrix, F_n, of order n, then F_n is nonderogatory only for $n = 1, 2, 3$. In these cases, one would then have to select A and B as polynomials in F_n. See [Davis 1979, pp. 33 and 232].

[80] The condition $\det(A) \neq 0$ is necessary and sufficent for the existence of $\log(A)$. See, e.g., [F. R. Gantmacher 1960, *Theory of Matrices*, Chap. 8, Sec. 8]. If the eigenvalues of A are all positive, then $\log(A)$ can be taken real. With t pure imaginary, $M = \exp(t \log(A))$ is then an instance of a "circular" matrix for which conj $(M) = M^{-1}$. See [de Bruijn and Szekeres 1955].

If A is nonsingular but is nondiagonalizable, considerable numerical difficulties may be experienced in computing $\log(A)$.

If A is singular and the zero eigenvalue is of index 1 (i.e., all its zero eigenvalues are associated with Jordan blocks of dimension 1), then it may be Jordanized as $A = Q$ diag $(O, B)Q^{-1}$, where O and B are square and B is nonsingular. In this case, one may define a fractional power of A by means of

$$A^t = Q \operatorname{diag}(O, B^t)Q^{-1},$$

where B^t is defined as before. In this case, A^t will satisfy the law of exponents and $A^1 = A$, but $A^0 = Q \operatorname{diag}(O, I)Q^{-1}$. Moreover, we will have $A^{-t} = (A^t)^{\div}$, where $\div$ designates the Moore–Penrose generalized matrix inverse.

If a zero eigenvalue of A is associated with a Jordan block of dimension higher than 1, then a fractional power satisfying the law of exponents may not be definable. (For example, the 2×2 matrix $A = [01; 00]$ has no square root.) See [Gantmacher 1960, Chap. 8, Secs. 6 and 7].

For a discussion of the matrix-Schroeder equation, see [Kuczma et al. 1968, Chap. 6, Sec. 4]. One might examine Theodorus from this point of view.

[81] If U is unitary and $\sigma > 1$, take $A = \sigma U$ and $B = -A$. Then the (spectral) norm $(A) = \sigma > 1$. It is easy to show that if norm $(v_0) > \sigma/(\sigma - 1)$, the orbit is unbounded. If norm $(v_0) \leq \sigma/(\sigma - 1)$, the orbit is bounded.

[82] See, e.g., [Davis 1979, p. 32].

[83] It is hard to avoid the observation that we are dealing here with a very rudimentary form of the mathematical problem of exterior ballistics or of missile guidance.

[84] My particular interest in this kind of theorem lies in its relationship to the Monte Carlo method of numerical analysis and to the philosophy and methodology of random and pseudorandom sequences.

Working in one dimension for simplicity of notation, it can be shown that a sequence x_n is equidistributed modulo 1 if and only if

$$\lim_{N \to \infty} \left(\frac{1}{N}\right) \sum_{k=1}^{k=N} f(x_k) = \int_0^1 f(x)dx,$$

for all bounded, Riemann integrable functions $f(x)$. Thus, once in posession of an equidistributed sequence, the value of an integral may be computed by sampling on that sequence. It is not necessary to sample "at random" – whatever that means; it suffices to sample on a deterministic equidistributed sequence of which many are known explicitly, and of which the sequence $x_n = (n\theta) = n\theta - [n\theta], \theta =$ irrational, is undoubtedly the simplest. By way of underlining this simplicity, such sequences are even constructible by ruler and compass methods of elementary geometry, as is the case with the Theodorus angles.

The numerical computation of integrals of dimension d presents severe problems when $d \geq 10$, say, and the Monte Carlo computational strategy is often selected. As an adjunct to this, a considerable literature has developed that provides explicit equidistributed sequences in higher-dimensional intervals or manifolds, and which obtains via the ideas of "discrepancy" specific equidistributed sequences for which the rate of convergence is particularly rapid. See, e.g, [Davis and Rabinowitz 1984, Chapt. 5.9].

Now if one is not aware of the notion of equidistribution and simply proceeds via probabilistic sampling, one needs to develop random sequences on the computer. Since traditionally defined notions of randomness are noncomputable, one settles for so-called pseudorandom or quasirandom sequences. Such sequences are deterministic and easily computable (they are theoretically periodic, but have an enormously long period) and have been "certified" as having passed a certain number of well-known statistical tests $T(1), \ldots, T(2)$, but may fail a number of tests $T(N+1), T(N+2), \ldots$. The first test $T(1)$ is usually that of equidistribution.

Scientific computation packages deliver pseudorandom sequences on call. These are generally linear congruential sequences $x_{n+1} = ax_n + b \pmod{m}$, for which "good" values of a, b, m are claimed, sometimes by theorem, and sometimes by experience. Other methods are also employed in systems-supplied random number generators.

Let us designate by J the numerical job (interpreted as a fairly extensive class of related jobs) to be accomplished by probabalistic methods. It should be clear from the case of approximate integration that, if you want to do the job J properly, all you need to require of the pseudorandom sequence is that it pass tests $T(r_1), \ldots, T(r_s)$, where the s-tuple $(r_1, r_2, \ldots, r_s)$ depends upon J.

Strangely, I think that there has been relatively little thematic or experiential discussion of (a) the independence or interrelation between the standard statistical tests $T(i)$ (for example: prove that $T(6)$ does or does not imply $T(7)$); (b) what tests T does a sequence need to pass in order to lead to a theoretically sound Monte Carlo method for job J.

It is clear from the theory of equidistribution that a sequence need only pass $T(1)$ to integrate properly. If it passes other tests as well, it is not clear a priori that this makes it a better sequence to use. It may actually have the reverse effect and slow down the converence from $O(N^{-(1-\epsilon)})$ to $O(N^{-1/2})$. This slowing down may be viewed as a price or a tradeoff that has to be paid for the "higher quality" randomness and hence the greater versatility of such a sequence.

Let me now take a big existential jump and postulate that for every numerical job J for which probabilistic methods are currently used, there is a set of tests $T(j_1), \ldots, T(j_{k(J)})$ that a deterministic sequence must pass in order to do J successfully. Furthermore, deterministic, computable sequences can be defined that pass these tests and that, consequently, can be used successfully as a basis for a Monte Carlo algorithm for J.

Call this hypothesis HJ. As stated, it is admittedly vague. (What is a test? What does it mean to "pass" such a test? What kind of computability are we talking about? Turing computability? Computability on a real machine within a human lifetime?) Perhaps the logical status of HJ might be analogous to Church's hypothesis that says that all effective computation can be formalized with the lambda calculus, or to the continuum hypothesis for which it is now known that you can believe it or not believe it, as you will.

What is the evidence in favor of HJ? The existence of equidistribution theory plus the fact that the random number generators delivered by computer systems seem, with some care, to do useful jobs.

But let me explore one consequence: if HJ were the case, then as far as computation is concerned, there is no need to introduce the notions of probability, randomness, etc. These notions might be a linguistic convenience, possibly even inspiration for algorithmic strategies, but not a necessity. Some of the metaphysical angst that now accompanies definitions of random sequences would be eliminated.

Passing from the world of formalized mathematics and computations to the world of physical and social events, we can say that to the extent that a theory of physics must have a computation as its real or potential endpoint, the language of probability is unnecessary. But of course, to say that probability is unnecessary in computation is not the same as asserting that the exterior world is deterministic. A major (unsolvable) problem is to delineate those parts of the extramathematical world that can be modelled by constructive, computable mathematics.

In summary, I think that the existence of the theory of equidistribution (which few mathematicians seem to be acquainted with) is one reason why the traditional expression of probability through the Kolmogoroff axiomatization is an inadequate expression of the mathematical realities.

Index

x P.9 & COVER

x P. 41
x P. 197
x P. 14
x p. 198
x p. 215
x P. 206
x P. 222
x Schroeder   P. 229
                    P. 43, 44

x P. 214   Explicate — to explain
P. 224   Apotheosized — exalted & deified
P. 207   Agapasticism

Tychism {Absolute chance is operative in the cosmos,
        {Variation (in evolution) may be purely fortuitous
        {chance is an objective reality
Synechism — continuity (of hypotheses) is of prime importance

Apercu — an insight; a summary